Studies in Economic Theory

Springer

Berlin
Heidelberg
New York
Hong Kong
London
Milan
Paris
Tokyo

Titles in the Series

M. A. Khan and *N. C. Yannelis* (Eds.)
Equilibrium Theory in Infinite Dimensional Spaces

C. D. Aliprantis, K.C. Border and *W. A. J. Luxemburg* (Eds.)
Positive Operators, Riesz Spaces, and Economics

D. G. Saari
Geometry of Voting

C. D. Aliprantis and *K. C. Border*
Infinite Dimensional Analysis

J.-P. Aubin
Dynamic Economic Theory

M. Kurz (Ed.)
Endogenous Economic Fluctuations

J.-F. Laslier
Tournament Solutions and Majority Voting

A. Alkan, C. D. Aliprantis and *N. C. Yannelis* (Eds.)
Theory and Applications

J. C. Moore
Mathematical Methods for Economic Theory 1

J. C. Moore
Mathematical Methods for Economic Theory 2

M. Majumdar, T. Mitra and *K. Nishimura*
Optimization and Chaos

K. K. Sieberg
Criminal Dilemmas

M. Florenzano and *C. Le Van*
Finite Dimensional Convexity and Optimization

K. Vind
Independence, Additivity, Uncertainty

T. Cason and *C. Noussair* (Eds.)
Advances in Experimental Markets

F. Aleskerov and *B. Monjardet*
Utility Maximization, Choice and Preference

Karl Vind

Independence, Additivity, Uncertainty

With Contributions by Birgit Grodal

With 10 Figures

Professor Karl Vind
University of Copenhagen
Institute of Economics
Studiestraede 6
DK 1455 Copenhagen K
Denmark

ISBN 3-540-41683-8 Springer-Verlag Berlin Heidelberg New York

Bibliographic information published by Die Deutsche Bibliothek
Die Deutsche Bibliothek lists this publication in the Deutsche Nationalbibliografie; detailed bibliographic data available in the internet at http.//dnb.ddb.de

Springer-Verlag Berlin Heidelberg New York
a member of BertelsmannSpringer Science + Business Media GmbH

http://www.springer.de

Printed in Germany

Cover design: Erich Kirchner, Heidelberg

SPIN 10797811 42/2202-5 4 3 2 1 0 – Printed on acid-free paper

Preface

The work on this book started many years ago as an attempt to simplify and unify some results usually taught in courses in mathematical economics. The economic interpretation of the results were representations of preferences as sums or integrals and the decomposition of preferences into utilities and probabilities. It later turned out that the approach taken in the earlier versions were also the proper approach in generalizing from preferences which were total preorders to preferences which were not total or transitive. The same mathematics would even in that situation give representations which were additive. It would also give decompositions where concepts of uncertainty appeared.

Early versions of some of the results appeared as Working Papers No. 135, 140, 150, and 176 from The Center for Research in Management Science, Berkeley.

A first version of chapters 2, 4, 6, 7, and 8 appeared 1969 with the title "*Mean Groupoids*" [177]. They are essentially unchanged - except for some notes especially in chapter 6. Another version appeared 1990 as [178]. Chapter 10 contains results from the same versions and from [181].

Chapter 11 by Birgit Grodal is based on [91] by Grodal and Jean-Francois Mertens. Chapters 11 and 12 - also by Birgit Grodal - contains the results from the earlier versions, but have been extended (by Karl Vind) to take into account the new corollaries of the results in the other chapters.

The realization at MSRI, Berkeley 1985-86 that the same mathematics could be used to get results for relations which were not total or transitive resulted in the papers [180, 179]. The results in these papers are included in and extended in this book. They were presented January 1987 in Oberwolfach, where I also heard Bewley's ideas about Knightian uncertainty. The importance of the results about not totally ordered function spaces for formalizing uncertainty became clear in conversations with Truman F. Bewley in Bonn in the summer of 1990.

The *Notes* gives references, open problems and a few theorems.

The *References* contains the references. No attempts have been made to make a complete bibliography of any of the fields touched upon in the book[1], but some attempts have been made to include references to papers and books which may be relevant to or extending the results in this book

The work on this book started in 1965, and parts of the results have since been used in courses and presented at seminars in mathematical economics at the University of Copenhagen, the University of California, Berkeley, Stanford University, etc. I have had very useful discussions with and comments from a large number of economists, mathematicians, and statisticians. I should like to mention in particular Gérard Debreu, Werner Fenchel, Birgit Grodal, and Søren Johansen.

Mansoor Hussain did a good job proofreading, the remaining mistakes were probably added after he finished.

My research has been supported by grants from the Social Science Research Council, Denmark, the Carlsberg Foundation, by grant from the Ford Foundation to the Graduate School of Business Administration, University of California, Berkeley, and by NSF grant 8120790. Many of the new results in the book have been obtained at the University of California, Berkeley, as a visitor to the Department of Economics (1962-63, 1964-66, 1970-71, 1981-82, 1983, 1990-91) or to Mathematical Sciences Research Institute (1985–86).

August 2002 Karl Vind

[1] For most of the fields see Peter Wakker's useful references [182].

Contents

[1]Letter of May, 1983 to Paul Samuelson, copy to me summer 2000.

III Relations on Measures 131

IV Integral Representations 147

List of Figures

1

Introduction

1.1 Economics

The most important contribution of this book to economic theory is to formalize a concept of uncertainty. Ideas about uncertainty and this concept's importance for economics go back at least to Keynes (1921) [106] and possibly Knight (1921) [110]. The ideas have been rejected with the argument that the concept could not be made precise and therefore not included in serious economic theory. To the extend that what is meant by uncertainty is that agents can not compare any two alternatives, uncertainty can be represented by real numbers just like utility, probability and preference.

The paradigm changes from:

A is preferred to B if the (expected) utility of A is larger than the (expected) utility of B

to:

A is preferred to B if the (expected) preference for A over B is larger than the (expected) uncertainty.

Only additive set functions appear in the representation theorems.

1.2 Statistics

No theory of uncertainty and its importance for the choices made by agents can be taken seriously without the relation to the foundation of statistics being made explicit. The foundation of statistics sketched here (chapter 18 page 243) gives a compromise between the Bayesian view that the prior knowledge about subsets of the parameter space is a total preorder representable by a probability measure and the other extreme that only the observations can say anything about where in the parameter space the right parameter is. The compromise is not only that there is uncertainty on the parameter space, but contains also the possibility that a point in the parameter space gives uncertainty on the outcome space. In this way also uncertainty about whether the parameter space is the right space can be formalized and expressed.

1.3 Mathematics

From a mathematical point of view this is a book about function spaces with an (pre)order relation. The results obtained for totally preordered spaces characterize sums of functions or integrals or measures as order preserving real functions defined on a set of functions with a total preorder. An independence assumption is necessary and sufficient for this representation. The generalization to relations not assumed to be total consists in observing that these results generalize to sets of functions with a relation for which the independence condition holds, but the relation does not have to be total or transitive. The additivity in the representation theorems is thus a consequence of the independence assumption and the assumption that the relation is total and transitive only gives special cases of the general theorems.

Most of the results in the book are based on a surprisingly simple result characterizing a subset of a product of n sets - an independence condition is equivalent to the product being isomorphic to (a subset of) $\mathbb{R}^n$, and the subset to an open half space. This result implies theorems characterizing totally preordered function spaces with an independence property via representation in the form of additive orderpreserving real functions, integrals being an obvious example.

The proof of the basic result on a subset of a product uses a result about ordered algebraic spaces (chapter 6). The Aczél-Fuchs theorem (theorem 13) page 55 proves, that there is an isomorphism between a commutative mean groupoid and a subset of the real numbers with the natural midpointoperation as the algebraic part of the structure.

This theorem is then used on a product of four sets where the independence condition implies first that a subset determines a total preorder and then that the Thomsen and the Reidemeister conditions known from the theory of nets will hold. This again implies - via the Aczél Fuchs theorem - that there is a representation, where an additive function is positive precisely on the subset. If the subset is the graph of a total preorder on a product of two sets, one obtains the well-known result that the Thomsen or the Reidemeister condition implies the existence of a representation of a total preorder by a sum of two functions.

By regarding function spaces as product spaces this representation can be used to get representation of both relations which are total and transitive and relations which only have the independence property. When the relation is total the functions space is a commutative mean groupoid on which the Aczél Fuchs theorem can be applied. When the relation is not total the product of the function space with itself is a commutative mean groupoid on which the Aczél Fuchs theorem can be applied.

In the case of relations which are not total and/or transitive representations in the form of sums or integrals which are positive if and only if the relation holds are obtained. For these representations decompositions into preferences, probability, and uncertainty are obtained by trivial rewritings.

1.4 Summary of results

Ch.	Structure	Assumptions
2	$(S, \succsim)$	Connected separable
4	$(\prod S_i, \succsim)$	c.3 (connected)
5	$(\prod S_i, Q)$	Independence
6	$(S, \succeq, \circ)$	Bisymmetry
7	$(X \times Y, \succsim)$	Independence
8	$\mathcal{G} \subset Y^X$, $(\mathcal{G}, \succsim)$	Independence
9	$\mathcal{G} \subset Y^X$, $(\mathcal{G}, Q)$ $\mathcal{G} \times \mathcal{H} \subset Y_1^X \times Y_2^X$, $(\mathcal{G} \times \mathcal{H}, \mathcal{P})$	Independence Independence
11	$(\mathcal{G}, \succsim)$	$+Y$ separable metric, topology t order topology $\subset t$
12	$(\mathcal{G}, \succsim)$	$(Y_A, \succsim_A)$ independent of A $(Y_A, \succsim_A, \circ_A)$ independent of $\mathcal{A}$ $X = \mathbb{R}$ Translation invariant

Main result	Theorem	Page
$(S,\succeq) \leftrightarrow ([0,1],\geqq)$	5	23
$(\prod S_i,\succeq) \longmapsto ([0,1],\geqq)$	10	37
$(S,\succsim), Q=\{s\succ s_0\}$	11	42
$(S,\succeq,\circ) \longmapsto (\mathbb{R}_1,\geqq,\circ)$	13	55
$F(x,y)=F_1(x)+F_2(y)$	19	74
$F(g,X)>F(h,X) \Longleftrightarrow g\succ h$; $F(g,\cdot)$ ADDITIVE	26	95
$F(g,X)>0 \Leftrightarrow g\in Q$; $F(g,\cdot)$ ADDITIVE $F(g,h,X)>0 \Leftrightarrow g\in P(h)$; $F(g,h,\cdot)$ ADDITIVE	36 Cor. 10	113 115
$F(g,A) = \int_A u(x,g(x))\,d\mu$	52	155
$F(g,A) =$ $\int_A \beta(x,\overline{u}(g(x)))\,d\mu$	56	171
$\int_A u(g(x))\,\alpha(x)\,d\mu$	57	174
$\int_A \overline{u}(g(t))\,e^{-\delta t}dt$	58	178

1.5 Applications

Additive functions, integrals and measures appear of course in many theories. In many of these applications it gives a deeper foundation to regard these functions as order preserving real functions defined on a space of functions with a given total preorder. This book gives results needed for this kind of representations. For many of the applications it is also very important to know that exactly the same mathematical results can be used to give representations also for the case, where the given relation on the function space is not total. This new knowledge is the main contribution of this book. The applications themselves are only mentioned in the notes or sketched. (In chapters 16, 17, and 18).

For some of the important applications special extra structure (the range of the functions may be a metric space or a commutative mean groupoid and the domain may be a topological space or even $\mathbb{R}$) can on the one hand give further results - for example discounted utility for preferences over time - but on the other hand the independence assumption may be too strong. If the functions are defined on a product space - for example in the foundation of statistics, where the basic set can be considered to be a product of a parameter space and an outcome space - it may be necessary to replace the independence assumption with two-level independence, where there is independence on functions on the parameter space, and given a value of the parameter there is independence on the outcome space. If the space where the functions are defined has a a total order - for example time - the independence condition may be relaxed in a different way (see chapter 17). So for many applications the results obtained in the representation theorems should be regarded as technical tools needed to obtain results under the weaker assumptions which can be formulated with the special structures on the range and the domain of the function spaces, and which may be more acceptable from the point of view of the applications.

Part I

Basic Mathematics

2

Totally preordered sets

2.1 Introduction

The central topic in this chapter is totally preordered sets. Section 2.2 gives conditions for the existence of order homomorphisms between totally preordered sets and subsets of the real numbers. Some topological concepts are introduced in section 2.3 page 18. A total order gives rise to a topology and the main results from section 2.2 are in section 2.4 given in topological terms.

Most of the contents of this chapter are well-known and can be found in many texts. It is therefore given here without proofs. Almost all results on totally ordered spaces are, however, proved.

2.2 Order relations

2.2.1 Basic concepts

The following seven axioms will be used in the definitions of order structures on a set S with a relation R.

Axiom		Definition
O1	Reflexive	aRa
O2	Antisymmetric	aRb and bRa imply $a = b$
O3	Total	aRb or bRa
O4	Transitive	aRb and bRc imply aRc
O5	Symmetric	aRb implies bRa
O6	Asymmetric	aRb implies not bRa
O7	Irreflexive	not aRa

Definition 1 (preorder) *A relation R is a preorder and denoted $\succsim$ if O1 and O4 hold.*

Definition 2 (order) *A preorder is an order if also O2 holds.*

Definition 3 (total) *A preorder (order) is total if O3 holds.*

Definition 4 (partial) *A preorder (order) is partial if O3 does not hold.*

Notation 1 *A preordered set will be denoted* $(S, \succsim)$. *An ordered set will be denoted* $(S, \succeq)$.

Definition 5 ($\sim$) *The relation* $\sim$ *on* S *can for any preorder* $\succsim$ *be defined by*

$$a \sim b \text{ if } a \succsim b \text{ and } b \succsim a$$

Remark 1 $\sim$ *is an equivalence relation (i.e. O1, O4, and O5 hold).*

Definition 6 (quotient set) *The set of equivalence classes (or the quotient set) is denoted* $S' = S/\sim$ *and is an ordered set with the order defined by* $a' \succeq b'$ *if* $a \succsim b$, *where* $a \in a'$ *and* $b \in b'$ $(a', b' \in S' \text{and } a, b \in S)$.

Definition 7 (canonical projection) *The projection* $\chi : S \to S'$ *defined by* $\chi(s) = s'$ *for* $s \in s'$ *will be called the canonical projection.*

Definition 8 ($\succ$) *The relation* $\succ$ *is defined on* S *by* $a \succ b$ *if* $a \succsim b$ *and not* $b \succsim a$. $\precsim$ $(\prec)$ *is defined by* $a \precsim b$ $(a \prec b)$ *if* $b \succsim a$ $(b \succ a)$.

Remark 2 $\succ$ *will except for O1 have the properties of* $\succsim$. $\succ$ *will be asymmetric and irreflexive (i.e. have properties O6 and O7).*

The following table summarizes the notation and the definitions and properties of relations

	O1	O2	O3	O4	O5	O6	O7
preorder	+			+			
order	+	+		+			
total			+				
partial			-				
$\succsim$	+			+			
$\sim$	+			+	+		
$\succeq$	+	+		+			
$\succ$	-			+		+	+

Definition 9 (maximal (minimal)) *An element* $a \in A \subset S$ *is a maximal (minimal) element in* A *if for all* $b \in A$, $b \succsim a$ $(a \succsim b)$ *implies* $b \sim a$.

Definition 10 (first (last)) *An element* $a \in A \subset S$ *is a first (last) element in the totally preordered set* $(A, \succsim)$ *if* $s \succsim a$ $(a \succsim s)$ *for all* $s \in A$.

Definition 11 (interval) *The set*

$$\{s \in S \,|\, b \succsim s \succsim a\} \text{ is denoted } [a, b]$$

and called the closed interval with a *as the initial point and* b *as the terminal point.*

$$]a, b[= \{s \in S \,|\, b \succ s \succ a\}$$

is called an open interval. We shall also call

$$\begin{gathered}]a, b] = \{s \in S \,|\, b \succsim s \succ a\} \\ [a, b[= \{s \in S \,|\, b \succ s \succsim a\} \\ S \\ [a, \rightarrow[= \{s \in S \,|\, s \succsim a\} \\ ]a, \rightarrow[= \{s \in S \,|\, s \succ a\} \\ ]\leftarrow, a] = \{s \in S \,|\, a \succsim s\} \\ \text{and} \\ ]\leftarrow, a[= \{s \in S \,|\, a \succ s\} \end{gathered}$$

intervals.

Definition 12 (gap) *A pair* (a, b) *with* $b \succ a$ *is called a gap in* S *if* $]a, b[$ *is empty.*

Definition 13 (minorant (majorant)) $a \in S$ *is called a minorant (majorant) of* $A \subset S$ *if* $[a, \rightarrow[\, \supset A$ $(]\leftarrow, a] \supset A)$.

Definition 14 (bounded) *If* A *has a minorant and a majorant it is bounded.*

$A \subset S$ can be considered a preordered set with the relation $\succsim'$ defined by $a \succsim' b$ if $a \succsim b$ for $a, b \in A$. If Oi $(i = 1, 2, 3, 4)$ holds for $\succsim$ it will also hold for $\succsim'$ and we shall usually denote $\succsim'$ by $\succsim$.

Definition 15 (supremum (infimum)) *A majorant a of A is a supremum (infimum) of A if for any majorant (minorant) a', $a' \succsim a$ ($a' \precsim a$).*

Definition 16 (sup (inf)) *The set of suprema (infima) of a set A will be denoted* $\sup A$ ($\inf A$)*. For an ordered set where a supremum and an (infimum) is unique* $\sup A$ ($\inf A$) *denotes an element and not a subset.*

Definition 17 (lattice) *An ordered set $(S, \succeq)$ is a lattice if* $\sup(a, b)$ *and* $\inf(a, b)$ *exist for any pair (a, b).*

Definition 18 (order-complete) *A preordered set is order-complete if for any bounded set A* $\sup A$ *and* $\inf A$ *exist.*

Definition 19 (dense) *A set A is dense in a preordered set $(S, \succsim)$ if for any $a, b \in S$, $a \succ b$, $]b, a[\neq \emptyset$ there exists a $c \in A$ such that $a \succ c \succ b$.*

Definition 20 (separable) *A totally preordered set S is separable if there exists a countable set A such that for any $a \succ b$ there exists a $c \in A$ such that $a \succsim c \succsim b$.*

If $(S, \succsim)$ and $(T, \succsim)$ are preordered sets and $f : S \to T$ is a function, we define f to be

Definition 21 (order preserving) *if $a_1 \succsim a_2 \Rightarrow f(a_1) \succsim f(a_2)$.*

Definition 22 (strictly monotonic) *if $a_1 \succ a_2 \Rightarrow f(a_1) \succ f(a_2)$.*

Definition 23 ((order) homomorphism) *if it is order preserving and strictly monotonic.*

Definition 24 ((order) isomorphism) *if it is bijective and f and f^{-1} are order preserving.*

Remark 3 *If both preorders are total and S is an ordered set, then strictly monotonic implies order preserving.*

Definition 25 (isomorphic) *$(S, \succsim)$ and $(T, \succsim)$ are isomorphic if there exists an isomorphism between them.*

Remark 4 *Isomorphic is an equivalence relation.*

Definition 26 (homomorphic) *$(S, \succsim)$ is homomorphic with $(T, \succsim)$ if there exists a homomorphism $f : S \to T$.*

In some interpretations the set S is a set of actions for an agent and the relation $\succsim$ is a preference relation for the agent on the set of actions. In this interpretation the following terminology is traditional.

Definition 27 (utility, utility function) *A homomorphism f for $(T, \succsim) = (\mathbb{R}, \geqq)$ is called a utility function, and the value of the function, $f(t)$, is called the utility of t.*

Remark 5 *Let $(S, \succsim)$ be a preordered set, $(S', \succeq)$ the ordered set of equivalence classes, and $\chi : (S, \succsim) \to (S', \succeq)$ the canonical projection.*

$$\begin{array}{ccc} (S, \succsim) & \overset{\chi}{\rightarrow} & (S', \succeq) \\ & \overset{f}{\searrow} & \updownarrow f' \\ & & (T, \succeq) \end{array}$$

If

$$f' : (S', \succeq) \leftrightarrow (T, \succeq)$$

is an isomorphism, then

$$f = f' \circ \chi : (S, \succsim) \to (T, \succeq)$$

is a homomorphism.

Example 1 *The set of all subsets of a set S denoted 2^S is partially ordered by $\subset$ (totally ordered if S has one element or is empty). Any subset is bounded by $[\emptyset, S]$, $\emptyset = \inf 2^S$, $S = \sup 2^S$, and 2^S is a lattice with $\inf(A, B) = A \cap B$ and $\sup(A, B) = A \cup B$.*

Example 2 *The sets $\mathbb{R}, \mathbb{Q}$, and $\mathbb{Z}$ are totally ordered by $\geqq$. $\mathbb{R}$ and $\mathbb{Z}$ but not $\mathbb{Q}$ are order-complete. $\mathbb{Q}$ is dense in $\mathbb{R}$, no proper subsets of $\mathbb{Z}$ are dense in $\mathbb{Z}$, but $\mathbb{Z}$ is separable. $\mathbb{R}$ and $\mathbb{Q}$ have no gaps, but $\mathbb{Z}$ has gaps. $\mathbb{R}, \mathbb{Q}$, and $\mathbb{Z}$ have no first and last elements.*

Definition 28 *If $(S, \succsim)$ and $(T, \succsim)$ are preordered sets we can define two different preorders on $S \times T$. First the product preorder*

$$(s,t) \succsim (s',t') \text{ if } s \succsim s' \text{ and } t \succsim t'$$

this is a partial preorder even if S and T are totally preordered sets (and have more than more than one elements each). Second the lexicographic preorder is defined by

$$(s,t) \succsim (s',t') \text{ if } s \succ s' \text{ or } (s \sim s' \text{ and } t \succsim t')$$

Definition 29 *If $(S_\alpha, \succsim_\alpha)_{\alpha \in A}$ are preordered sets we can generalize the product preorder by defining*

$$(s_\alpha)_{\alpha \in A} \succsim (s'_\alpha)_{\alpha \in A} \text{ if } s_\alpha \succsim s'_\alpha \text{ for all } \alpha \in A$$

or if A is a well-ordereded set[1] *the lexicographic preorder by defining ($\nsim$ defined by not $\sim$)*

$$(s_\alpha)_{\alpha \in A} \succsim (s'_\alpha)_{\alpha \in A} \text{ if } s_\alpha \sim s'_\alpha \text{ for all } \alpha \in A$$
$$\text{or}$$
$$s_\beta \succsim s'_\beta \text{ where } \beta \text{ is the first element in } A \text{ with } s_\alpha \nsim s'_\alpha$$

The first relation defined in definition 29 is often used for function spaces. $\mathcal{G} \subset Y^X = \{g \,|\, g : X \to Y\}, (Y, \succsim)$. $(\mathcal{G}, \succsim)$ is then defined by $g_1 \succsim g_2$ if $g_1(x) \succsim g_2(x)$ for all $x \in X$.

2.2.2 Completion

If $(S', \succeq)$ is a totally ordered set and S' is not order-complete we can define a totally ordered order-complete set $(S, \succeq)$ such that $(S', \succeq)$ is isomorphic to a dense subspace of $(S, \succeq)$. To prove this we need a definition.

Definition 30 (cut) *A cut in $(S', \succeq)$ is a pair (A, B) of non-empty subsets of S', where A is the set of all minorants of B, and B the set of all majorants of A. For any $s \in S'$ $(]\leftarrow, s], [s, \rightarrow[)$ is a cut (determined by s).*

---

[1] A totally ordered set where every non-empty subset contains a first element.

Lemma 1 *Let (A, B) be a cut in S', then $A \cup B = S'$ and $A \cap B = \emptyset$ or $A \cap B = \{s\}$ for some $s \in S'$. $(S', \succeq)$ is order-complete if and only if every cut is determined by an element from S'.*

Proof. Let $s \in S'$, if $s \succeq b$ for some $b \in B$, then $s \succeq b \succeq a$ for all $a \in A$ and $b \in B$, otherwise $s \prec b$ for all $b \in B$ and $s \in A$ so $A \cup B = S'$. If s_1, s_2 are both elements in $A \cap B$ then $s_1 \succeq s_2$ and $s_2 \succeq s_1$ so $s_1 = s_2$.

If $(S', \succeq)$ is order-complete $\sup A = \inf B = s \in S'$ for any cut (A, B). Let A be a set with a majorant and B the set of all majorants of A. If all cuts are of the form $(]\leftarrow, s]), ([s, \rightarrow[)$, then $B = [s, \rightarrow[$ and $s = \sup A$. If A has a minorant the existence of $\inf A$ follows dually ■

In chapter 6 the following definitions shall be needed.

Definition 31 ((A, B)) *For any pair of subsets (A', B') of S', we can define (A, B) by*

$$A = \{s \in S' \,|\, s \preceq b \text{ for all } b \in B'\}$$
and
$$B = \{s \in S' \,|\, s \succeq a \text{ for all a} \in A'\}.$$

Definition 32 (**(A', B') determines cut**) *If (A, B) is a cut, it is a cut determined by (A', B').*

It is obvious that if (A', B') determines a cut then $A' \succeq B'$ and $A' \cap B' = \emptyset$ or $\{c\}$.

An element c determines the same cut as the pair $(\{c\}, \{c\})$. Two pairs (A', B') and (A'', B'') determining cuts determine the same cut if and only if $A' \preceq B''$ and $A'' \preceq B'$. This constitutes a relation between pairs of subsets. This relation is obviously an equivalence relation. If a pair (A', B') determines a cut in S', (A', B') is a cut in $A' \cup B'$. If (A', B') and (A'', B'') determines cuts and $A' \subset A''$ and $B' \subset B''$, then they determine the same cut.

Definition 33 (**completion**) *Define now S as the set of cuts in S' and the relation $\succeq$ on S by $(A, B) \succeq (A', B')$ if $A \subset A'$.*

Theorem 1 *$(S, \succeq)$ is a totally ordered order-complete set. $(S', \succeq)$ is isomorphic to a dense subspace of $(S, \succeq)$ consisting of cuts determined by elements from S'.*

Proof. S is obviously totally ordered, and if $S_1 \subset S$ is bounded the set $B_0 = \bigcap_{(A,B)\in S_1} B$ and the set A_0 of all minorants of B_0 form a cut (A_0, B_0) in S'. $(A_0, B_0) = \sup S_1$. $\inf S_1$ can be found analogously and $(S, \succeq)$ has been shown to be order-complete. It follows easily from the definitions that if $a \prec b$, $a, b \in S$, and $]a, b[\neq \emptyset$ there exists a $c' \in S'$ such that $a \prec c \prec b$ where c is the cut determined by c'. It is also obvious that the function $c' \longmapsto c$ is an isomorphism between S' and the cuts determined by elements form S' ■

Remark 6 *Theorem 1 and lemma 1 shows that nothing is added by going through the process of taking cuts in S. The set of cuts in the set of cuts in S' is isomorphic to $(S, \succeq)$.*

2.2.3 Representation

Order isomorphisms give an equivalence relation among ordered sets. We are particularly interested in the equivalence classes containing subsets of $\mathbb{R}$ and $\mathbb{Q}$. We are in other words interested in finding conditions for the existence of an order isomorphism between $(S, \succeq)$ and $(\mathbb{R}_1, \geqq)$ where $\mathbb{R}_1 \subset \mathbb{R}$.

Theorem 2 *Let $(S, \succeq)$ be an ordered set*

$$(S, \succeq) \text{ is isomorphic to } ([0,1] \cap \mathbb{Q}, \geqq)$$

$$\Leftrightarrow$$

S is countable, totally ordered,
without gaps and has a first and last element.

Proof. The conditions are invariant under order isomorphisms and hold trivially for $([0,1] \cap \mathbb{Q}, \geqq)$, so they are obviously necessary. Let

$$S \setminus \{s_0, s_0'\} = \{s_1, s_2, \cdots\}$$

where s_0, (s_0') is the first (last) element and

$$]0,1[\cap \mathbb{Q} = \{r_1, r_2, \cdots\}$$

Define now $f : S \to \mathbb{Q}$ by $f(s_0) = 0$, $f(s_0') = 1$, $f(s_i) = r_j$ where r_j is the first element in $]0,1[\cap \mathbb{Q}$ not already chosen, such that the restriction of f to $\{s_1, s_2, \cdots, s_i\}$ is strictly monotonic. Both S and $[0,1] \cap \mathbb{Q}$ are without gaps and this implies that all r_j will be chosen and that the choice of r_j is always possible. The order is total and f is strictly monotonic and bijective so f^{-1} is strictly monotonic and f is an isomorphism ■

Theorem 3 (real representation) *Let $(S, \succeq)$ be a totally ordered infinite set*

$$(S, \succeq) \text{ is isomorphic to } ([0,1], \geqq)$$
$$\Leftrightarrow$$
S is without gaps, order-complete,
separable and has a first and last element.

Proof. The conditions are invariant under order isomorphisms and hold trivially for closed intervals of $\mathbb{R}$, so the only if part of the theorem is trivial. S is separable and has no gaps, so there exists a countable dense subset $(A, \succeq)$ containing the first and last elements. The conditions for theorem 2 now hold for $(A, \succeq)$, so there exists an isomorphism $f : A \to [0,1] \cap \mathbb{Q}$, the extension (also denoted f) $f : S \to [0,1]$ defined by $f(s) = \sup\{f(a) \,|a \in A, s \succeq a\}$ is clearly strictly monotonic and thus injective. For any $r \in]0,1]$, $r = f(s)$ where $s = \sup\{a \in A \,|r \geqq f(a)\}$ so f is surjective. f is bijective and order preserving so f^{-1} is order preserving and f is an isomorphism ■

It is obvious how theorems 2 and 3 can be reformulated for the case where S has no first or last element. If S has gaps or if $(S, \succeq)$ in theorem 3 is not order-complete, the construction from the proofs can still be applied and $(S, \succeq)$ will be isomorphic to $(f(S), \geqq)$ where $f(S) \subset [0,1]$.

Corollary 1 *Let $(S, \succsim)$ be a totally preordered set with more than one equivalence class*

$$(S, \succsim) \text{ is homomorphic with } ([0,1], \geqq)$$
$$\Leftrightarrow$$
S is without gaps, order-complete,
separable and has a first and last element.

Proof. If a totally preordered set $(S, \succsim)$ has the properties used in theorems 2 and 3 (without gaps, order-complete, separable and first and last element), then the totally ordered set $(S/\sim, \succeq)$ has the same properties, and theorem 3 combined with remark 5 page 13 gives the result ■

2.3 Topological concepts

Definition 34 (topology) *S is an arbitrary set and $t \subset 2^S$. t is a topology if $S, \emptyset \in t$ and arbitrary unions and finite intersections of subsets from t are in t.*

Definition 35 (topological space) *(S, t) where t is a topology on S is a topological space.*

Definition 36 (open) *The elements from t are called open sets.*

Definition 37 (closed) *The complement of an open set is closed.*

Remark 7 *S and $\emptyset$ are both open and closed.*

Definition 38 (interior) *The union of the open subsets of a set A is the interior of A, denoted $\mathring{A}$.*

Definition 39 (closure) *The complement of the interior of the complement of A, is the closure of A, denoted $\overline{A}$.*

Definition 40 (dense) *If $\overline{B} \supset A, B$ is dense in A.*

Definition 41 (separable) *(S, t) is separable if S contains a countable set dense in S.*

Definition 42 (connected) *S is connected if $A \cup B = S, A \cap B = \emptyset$ for $A, B \in t$ implies A or $B = S$.*

Remark 8 *The only subsets of a connected set which are both closed and open are therefore S and $\emptyset$.*

Definition 43 (interior point) *If $x \in \mathring{A}$, x is an interior point of A.*

Definition 44 (neighorhood) *If $x \in \mathring{A}$, $\left(B \subset \mathring{A}\right)$ A is a neighborhood of x, (B).*

Definition 45 (base) *A system of subsets $b \subset t$ is a base for the topology t if for all $A \in t, A = \bigcup_{i \in I} A_i$ with $A_i \in b$.*

Definition 46 (subbase) *$s \subset b \subset t$ is a subbase (for the topology t with the base b) if for all $A \in b, A = \bigcap_{i=1}^{n} A_i$ with $A_i \in s$.*

Remark 9 *A base or a subbase determines a topology.*

Definition 47 (finer) *t_1 is finer than t_2 if $t_1 \supset t_2$.*

Definition 48 (coarser) *t_1 is coarser than t_2 if $t_1 \subset t_2$.*

Remark 10 *$t = 2^S$ is the finest topology on S, and $\{\emptyset, S\}$ is the coarsest. If $b_1 \supset b_2$ ($s_1 \supset s_2$) for bases (subbases) for topologies t_1 and t_2 then $t_1 \supset t_2$ and t_1 is finer than t_2.*

Definition 49 (continuous) *If S and T are topological spaces and $f : S \to T$, then f is continuous if $f^{-1}(A)$ is open in S for all sets A open in T.*

Remark 11 *It is enough to check this condition for A from a base or a subbase.*

Definition 50 (homeomorphism) *A continuous bijection such that f^{-1} is continuous is a topological isomorphism or a homeomorphism.*

Definition 51 (homeomorphic) *Two topological spaces are homeomorphic if there exists a homeomorphism between them.*

Definition 52 (relative topology) *If A is a subset of a topological space (S, t) then $t_1 = \{A \cap B \,|\, B \in t\}$ is a topology on A called the relative topology (induced by t).*

Definition 53 (cover) *$(A_i)_{i \in I}$ is a cover of A if $\bigcup A_i \supset A$.*

Definition 54 (open cover) *$(A_i)_{i \in I}$ is an open cover of A if $\bigcup A_i \supset A$, and $A_i \in t$ for $i \in I$.*

Definition 55 (compact) *A set A is compact if for any open cover $(A_i)_{i \in I}$ of A there exists a finite set $I_0 \subset I$ such that $(A_i)_{i \in I_0}$ is a (sub)cover of A.*

Definition 56 (converge $\to$) *A sequence of elements from (S, t), $(s_i) = (s_1, s_2, \cdots)$ is said to converge to s_0 ($s_i \to s_0$) if every neighborhood of s_0 contains all except a finite number of the elements from the sequence.*

Notation 2 ($\uparrow, \downarrow$) *When $(S, t, \succsim)$ is a preordered set the notation $s_n \uparrow s_0$ ($s_n \downarrow s_0$) will be used instead of $s_n \to s_0$ if (s_n) is an increasing ($s_{n+1} \succsim s_n$) (decreasing ($s_{n+1} \precsim s_n$)) sequence.*

Definition 57 (product topology) *If $(S_i, t_i)_{i \in I}$ are topological spaces the product topology on $S = \prod_{i \in I} S_i$ is defined by choosing a base of the form $\left(\prod_{i \in I} B_i\right)$ where $B_i \in t_i$ and $B_i = S_i$ for all except a finite number of $i \in I$.*

Remark 12 *If $(S_i, t_i)_{i \in I}$ are connected then (S, t) is connected. If $(S_i, t_i)_{i \in I}$ are separable and I has at most countably many elements, then (S, t) is separable.*

Remark 13 *$(s_{ni})_{i \in I}$ converges to $(s_i)_{i \in I}$ if and only if $s_{ni} \to s_i$ for all $i \in I$.*

Definition 58 (metric) *A function $d : S \times S \to \mathbb{R}_+$ is a metric on S if $d(x, y) = d(y, x)$, $d(x, y) = 0$ if and only if $x = y$, and $d(x, y) + d(y, z) \geqq d(x, z)$.*

Definition 59 (metric space) *(S, d) is a metric space.*

Definition 60 (metric topology) *Sets of the form*

$$\{s \in S \,|d(x, y) < \varepsilon\}, \varepsilon > 0, \ y \in S$$

form a base for the metric topology.

2.4 The order topology

Definition 61 (order topology) *When $(S, \succsim)$ is a totally preordered set the order topology is defined as the coarsest topology on S such that open intervals are open sets.*

S and sets of the form $]a, \to[$ and $]\leftarrow, a[$ form a subbase for this topology. This subbase and sets of the form $]a, b[$ form a base and open sets have the form $\bigcup_{i \in I}]a_i, b_i[$. A closed interval is a closed set, but a closed set is not necessarily the union of closed intervals.

The relative topology of a subset A of S may be strictly finer than the order topology of the subset. It is easy to check that the relative topology on $A \subset S$ is equal to the order topology if and only if for all $s \in S$ either $s \sim a$ for some $a \in A$, $s \succ a$ for all

$a \in A, a \succ s$ for all $a \in A, b \succ s \succ a$, where (a, b) is a gap in A, or $A \cap]\leftarrow, s[$ and $A \cap]s, \rightarrow[$ have no first and last elements respectively. The only excluded possibilities are that $A \cap]\leftarrow, s[$ has a last element and $A \cap]s, \rightarrow[$ is not empty and has no first element and the dual.

If a preordered set is separable it is also separable in the order topology. A is dense in $(S, \succsim)$ if and only if it is dense in the topological space S.

If S is separable in the order topology and has at most countably many gaps, then $(S, \succsim)$ is separable. This means that we do not have to specify in which sense the concepts dense or separable are used for an ordered set with at most countably many gaps. A set with more than countably many gaps may be separable in the order topology but can not be separable in the order.

Example 3 *Let $S = [0,1] \times \{0,1\}$ and let $\succeq$ be the lexicographic order. $(\mathbb{Q} \cap [0,1]) \times \{0,1\}$ is dense so S is topologically separable, but S is not separable in the order.*

Theorem 4 *Let $(S, \succsim)$ be a totally preordered set*

$$(S, \succsim) \text{ is order-complete and has a first and last element}$$
$$\Leftrightarrow$$
$$(S, \succsim) \text{ is compact in the order topology.}$$

Proof. To prove that $\sup A$ exists for any $A \subset S$ when S is compact, assume that $\sup A$ does not exist, and denote by $B = \bigcup]b, \rightarrow[$ where the union is over all majorant of A, then $\cup_{a \in A}]\leftarrow, a[\cup B$ is an open cover of the compact S without a finite subcover. This gives a contradiction. $\inf A$ exists analogously. $\sup S$ and $\inf S$ consists of the last and first elements.

To prove compactness let $(A_i)_{i \in I}$ be an open cover of $S = [a, b]$, and let c_0 be the supremum of the c's for which $[a, c]$ can be covered by a finite set of A_i's. The supremum exists because of order-completeness and because a is in the set. If (c_0, b_0) is a gap for some $b_0 \succ c_0$, then $c_0 \in A_i$ and $b_0 \in A_j$ for some A_i, A_j, and the finite subcover could be extended to a finite subcover of also b_0. If no such gap exist and $c_0 \prec b$ the finite subcover can be extended to a finite subcover of a neighborhood of c_0 just by including any set A_i with $c_0 \in A_i$. This is a contradiction, so $c_0 \sim b$, and S is compact ■

To translate theorem 3 page 17 we shall also need

Lemma 2 *Let $(S, \succsim)$ be a totally preordered set*

$$(S, \succsim) \text{ has no gaps and is order-complete}$$
$$\Leftrightarrow$$
$$(S, \succsim) \text{ is connected in the order topology.}$$

Proof. Assume S connected, if (a, b) is a gap $]\leftarrow, a] =]\leftarrow, b[$ are both open and closed, so A connected implies that S has no gaps. Define the closed set $B \neq \emptyset, S$ by $B = \{s \in S \,|\, s \succsim a, \text{ for all } a \in A\}$ where A is a non-empty bounded set without supremum. $B = \bigcup_{b \in B} \{s \in S \,|\, s \succ b\}$ and therefore also open, again contradicting connectedness.

To prove S connected, let A be a set such that $A \neq \emptyset, S$. Let $a \in A$ and $b \notin A$ and for example $b \succ a$ and let $B = \bigcup [a, c]$ where the union is over the c's for which $b \succ c$ and $[a, c] \subset A$. $a \in B$ so $B \neq \emptyset$. $B \subset [a, b]$ so $d = \sup B$ exists. Any neighborhood of d contains a point of $B \subset A$ and of $S \backslash A$, so if A were both open and closed we should have $d \in A \cap (S \backslash A) = \emptyset$ ■

Lemma 3 *Let $(S, \succeq)$ and $(S', \succeq)$ be totally ordered spaces. An order isomorphism*

$$f : (S, \succeq) \leftrightarrow (S', \succeq)$$

is a homeomorphism. Let $(S, \succsim)$ be a totally preordered set and $(S', \succeq)$ a totally ordered space. A surjective order homomorphism

$$g : (S, \succsim) \rightarrow (S', \succeq)$$

is continuous.

Proof. Sets of the form

$$\begin{array}{ll} \{s \in S \,|\, s \succ a\} & \{s \in S \,|\, s \prec a\} \\ \{s' \in S' \,|\, s' \succ a'\} & \{s' \in S' \,|\, s' \prec a'\} \end{array}$$

form subbases for the order topologies so f and f^{-1} are continuous and f is bijective by definition of order isomorphism. g is continuous by an analogous argument ■

Definition 62 ($\succsim$ continuous) *Let $(S, t, \succsim)$ be a totally preordered topological space. $\succsim$ is continuous if the order topology is coarser than t.*

2.5 Representation

It is now easy to prove the main result of this chapter

Theorem 5 (real representation) *Let $(S, \succeq)$ be a totally ordered space*

$$(S, \succeq) \text{ is order isomorphic and homeomorphic to } ([0,1], \geqq)$$
$$\Leftrightarrow$$
$$S \text{ is connected, separable}$$
$$\text{and has a first and a last element.}$$

Proof. The only if part of theorem is trivial. Connected implies (lemma 2) that S has no gaps and is order-complete so theorem 3 page 17 gives the order isomorphism. Lemma 3 page 22 says that the order isomorphism is a homeomorphism ∎

Corollary 2 (existence of utility function) *Let $(S, t, \succsim)$ be a totally preordered topological space. If S is separable and connected in t, and $\succsim$ is continuous, then there exists a real continuous order homomorphism on S.*

Proof. S separable and connected in t implies S separable and connected in the coarser order topology, this implies $(S/\sim, \succeq)$ separable and connected in its order topology and lemma 3 and theorem 5 or the obvious modification for sets without first and/or last elements give the homomorphism ∎

It is obvious how theorem 5 can be reformulated for sets without first or last elements. If S is not connected the construction from the proof of theorem 2 page 16 will still give an order homomorphism with a subset of $[0, 1]$. The specific construction will imply that the order isomorphism is also continuous.

2.6 Notes

2.6.1 Basic concepts

The concepts of set, element, function, etc., which are assumed known in this book, and the understanding that it is possible "de faire dériver presque toute la mathématique actuelle d'une source unique, la Théorie des Ensembles" (Bourbaki (1939-) [33], Première Partie, Livre I (1960 chapter I-II page 4)) goes back to G.

Cantor. (Bourbaki (1939 -) [33] (Théorie des Ensembles), Fraenkel (1961) [78], Hausdorff (1914 og 1967) [94], Kamke (1950) [102] and Sierpinski (1958) [166] are comprehensive works in this field and discuss cardinal numbers, well-ordered sets, ordinal numbers etc.

[33] also contains a treatment of the general concepts of a structure and related concepts like homomorphism, homomorphic, isomorphic etc. The various concepts of structure introduced in this book (order, topology, mean groupoid), homomophism and isomorphism (order isomorphism, homeomorphism) are all special cases of these general concepts.

2.6.2 Ordered sets

The concept of a total order goes as far back as the concepts of time or number. The idea of a cut in a totally ordered set is due to Dedekind, and leads to one way of constructing the real numbers from the rational numbers.

The inclusion relation among subsets of a set is the first important example of an order which is not total. A systematic study of set with only an order structure started around 1900. The theory of lattices has been studied extensively since 1930.

2.6.3 Topology and order topology

The isolation of the concept of a topological space is due to Hausdorff (1914) [94], following the definition of a metric space by Fréchet in 1906. An order topology on an ordered space is another simple example of a topology, and many textbooks in topology (e.g. Kelley (1955) [104]) use order topologies as examples. Kowalsky (1965) [117] also uses order topologies as examples and gives in this way a systematic theory of the order topological properties of totally ordered sets. Many of the proofs in section 2.4 page 20 are versions of the proofs in [117]. Most of the theorems go back to Cantor or Dedekind.

2.6.4 Ordered topological spaces

Corollary 1 page 17 was first proved by Eilenberg (1941) [67] and introduced in economic theory by Debreu (1954) [50] to prove the existence of a real representation of preference ordering.

2.6.5 Lexicographic orders

If a totally ordered set is not separable we can get representations theorems using lexicographically ordered products of subsets of the real numbers. Results of this type goes back to at least 1914 (Hausdorff). Chipman (1960) [44] has argued for the importance for economic theory of this representation of a totally ordered set. Beardon et al. (2002) [17] have even classified totally ordered sets for which no real representation exist.

2.6.6 Removing gaps

Debreu (1964) [53] and Bowen (1968) [34] show that if there exists an order isomorphism $g : (S, \succeq) \to (\mathbb{R}_1, \geqq)$, where $\mathbb{R}_1 \subset \mathbb{R}$, then there exists a continuous $g' : (S, \succeq) \to (\mathbb{R}_2, \geqq)$ which is also an order isomorphism. This construction is not needed to generalize theorem 5 page 23 to the case where S is not connected, because the order isomorphism constructed in the proof of theorem 3 page 17 is automatically continuous even if S has gaps or is not order-complete.

2.6.7 Further results

The exercises in Bourbaki [33] *General topology* (English translation 1966)[2] contain many results on order topologies. Of the results not used in this book may be mentioned

1. There exists totally ordered connected sets with abitrary cardinality.

2. Given a topological space (X, t) one can find conditions for the existence of a total order on X such that t is the order topology.

3. A totally ordered set with the order topology is completely normal.

---

[2] See for example Chapter I § 1 ex. 2 a, § 2 ex 2,4,5, § 3 ex 8,9, § 11 ex 10, Chapter IV § 2 ex 6-17, §4 ex 7, §5 ex 1-6, 18, §6 ex 10 and Chapter IX §4 ex 4,5.

4. Debreu (1972) [55], Mas-Colell (1985) [136] chapter 2 and many others have studied the foundation of the classical assumption that utility functions are differentiable. See also Chipman Hurwicz Richter Sonnenschein (1971) [43]

5. Many results on real valued functions can be extended to functions with values in a totally ordered set.

These last results and results like theorem 4 in Debreu (1968) [54] and (IV) in appendix A page 182 in Hildenbrand (1970) [96] limit the importance of the results in this chapter. Instead of getting a real continuous order homomorphism from theorem 5 page 23 and then use known properties of real continuous functions, one could in many cases get the final result directly from the preorder without the restriction caused by introducing assumptions implying the existence of a real order homomorphism.

More results and references can be found in Maurice (1965) [138] and Birkhoff (1967) [26] chapter X.

3

Preferences and preference functions

3.1 Introduction

Chapter 2 treats totally ordered sets and gives representation theorems. Similar theorems for just relations - not assumed to be total - are trivial, but are convenient to have, because the main results in this book give conditions for particular additive representation.

3.2 Representations and representation theorems

The following representations and representation theorems will be of interest.

Let X and Y be arbitrary sets and $\mathcal{P}$ a relation on $X \times Y$.

Definition 63 (graph) $graph\mathcal{P} = \{(x, y) \,|y \in \mathcal{P}(x)\}$

Definition 64 (preference function) $v : Y \times X \to \mathbb{R}$ *is a preference function representing* $\mathcal{P} : X \to 2^Y$ *if*

$$v(y, x) > 0 \Longleftrightarrow y \in \mathcal{P}(x).$$

In many cases will X and Y be the same set, but in some applications can Y be X_i, where $X = \prod X_i$, so we formulate the results for the general case.

First we will however present an earlier result as a special case

Theorem 6 *Let X be a connected subset of $\mathbb{R}^\ell$ and $\succsim$ a continuous total preorder on X.*

$\mathcal{P}$ (defined by $\mathcal{P}(x') = \{x \in X | x \succ x'\}$) can be represented by $v(y, x) = u(y) - u(x)$, where $u : X \to \mathbb{R}$ is a continuous function.

Proof. Trivial given corollary 2 page 23 ∎

Remark 14 *Preference functions exist in general, for example*

$$v(y,x) = 1 \text{ for } y \in \mathcal{P}(x) \text{ and } v(y,x) = 0 \text{ for } y \notin \mathcal{P}(x).$$

Remark 15 *If v is a preference function for $\mathcal{P}$ and $f : \mathbb{R} \to \mathbb{R}$ has the property $f(x) > 0 \iff x > 0$, then $f \circ v$ is also a preference function.*

Remark 16 *Let X and Y be metric spaces with the metric both on $X \times Y$ and on Y denoted d.*

Define $v_1 : Y \times X \to \mathbb{R}$ and $v_2 : Y \times X \to \mathbb{R}$ by

$$v_1(y,x) = d((x,y), X \times Y \backslash graph\mathcal{P})$$
and
$$v_2(y,x) = d(y, X \backslash \mathcal{P}(x)) - d(y, \mathcal{P}(x))$$

Lemma 4 *If $graph\mathcal{P}$ is open, v_1 is a preference function. If $\mathcal{P}(x)$ is open for all x, v_2 is a preference function.*

Proof. Trivial ■

Let X and Y be affine metric spaces, where there is the relation between the two structures that

$$\lambda N(y_1, \varepsilon_1) + (1-\lambda) N(y_2, \varepsilon_2) =$$
$$N(\lambda y_1 + (1-\lambda) y_2, \lambda \varepsilon_1 + (1-\lambda) \varepsilon_2)$$

where $N(y, \varepsilon)$ is the $\varepsilon-$ball around y.

Lemma 5 *If $\mathcal{P}(x)$ is open and convex for all x, then $v_2(\cdot, x)$ is concave.*

Proof. Let $y^1 \in \mathcal{P}(x)$ and $y^2 \in \mathcal{P}(x)$, then open balls with radius $v_2(y^1,x)$ and $v_2(y^2,x)$ and centers in y^1 and y^2 are contained in $\mathcal{P}(x)$. $\mathcal{P}(x)$ is convex, so the ball with center in $\lambda y^1 + (1-\lambda)y^2$ and radius $\lambda v_2(y^1,x) + (1-\lambda)v_2(y^2,x)$ is contained in $\mathcal{P}(x)$, and $v_2(\lambda y^1 + (1-\lambda)y^2, x) \geq \lambda v_2(y^1,x) + (1-\lambda)v_2(y^2,x)$, implying that v_2 is concave in this case.

Let $y^1 \notin \mathcal{P}(x)$ and $y^2 \notin \mathcal{P}(x)$ and let $x^1 \in \overline{\mathcal{P}(x)}$ and $x^2 \in \overline{\mathcal{P}(x)}$ be points such that $v_2(y^1,x) = -d(y^1,x^1)$ and $v_2(y^2,x) = -d(y^2,x^2)$. $\mathcal{P}(x)$ is convex, so is $\lambda x^1 + (1-\lambda)x^2 \in \overline{\mathcal{P}(x)}$ and $v_2(\lambda y^1 + x, (1-\lambda)y^2) \geq -d(\lambda y^1 + (1-\lambda)y^2, \lambda x^1 + (1-\lambda)y^2) \geq \lambda v_2(y^1,x) + (1-\lambda)v_2(y^2,x)$. (Consider the smallest balls around

y^1 and y^2 containing x^1 and x^2. The convex combination of these balls will then contain the convex combination of x^1 and x^2). So v_2 is also concave in this case.

A similar construction covers the last case ∎

Lemma 6 *v_1 is continuous. If $\overline{\mathcal{P}}$ and $\mathcal{P}^c$ (defined by $\overline{\mathcal{P}}(x) = \overline{\mathcal{P}(x)}$ and $\mathcal{P}^c(x) = Y \backslash \mathcal{P}(x)$) are continuous*[1] *and Y compact, then v_2 is continuous.*

Proof. Follows directly from the maximum theorem, see for example Hildenbrand (1974) [97] or Debreu (1959) [51] ∎

We can now restate these results for the case $X \subset \mathbb{R}^m, Y \subset \mathbb{R}^n, Y$ compact and $\mathcal{P} : X \to 2^Y \setminus \{\emptyset\}$.

Theorem 7 *If graph$\mathcal{P}$ is open, a continuous representation exist. If $\overline{\mathcal{P}}$ and $\mathcal{P}^c$ are continuous and $\mathcal{P}(x)$ is open and convex for all $x \in X$, then there exists a continuous representation with $v(\cdot, x)$ concave.*

Proof. Contained in the previous lemmas ∎

Remark 17 *$\mathcal{P}^c$ upper hemicontinuous is equivalent to graph$\mathcal{P}$ open. Continuity of $\overline{\mathcal{P}}$ and $\mathcal{P}^c$ is therefore stronger than the assumption that graph$\mathcal{P}$ is open.*

Example 4 *The neat but of course not realistic preference function in $\mathbb{R}^n$*

$$v(y, x) = xy - xx$$

was given by Michelsen (1997) [139]. The unique demand is the point on the budget line closest to the origin. The preferred set is the set of points more expensive than x at the price $p = x$

3.3 Notes

In economic theory before 1970 some equivalence theorems have been stated and proved without the assumption that consumers had total and transitive preferences. The modern development where

---

[1] The continuity concepts for correspondences used here can be found in Hildenbrand (1974) [97]

also existence was proved without these assumptions started 1971, Sonnenschein (1971) [169], Mas-Colell (1974) [135] and Shafer Sonnenschein (1975) [164].

The microeconomic theory of the existence of and relations between preferences, and demand functions etc. has not been fully developed yet but it has been studied by among others Kihlstrom Sonnenschein (1976) [108], Kim, Richter (1986) [109], Michelsen (1997) [139], Quah (2000) [150], Epstein (1987) [69], and Shafer (1974) [163].

4

Totally preordered product sets

4.1 Introduction

The study of totally preordered product sets starts in this chapter. Section 4.3 contains definitions and some results concerning topological properties of totally preordered product sets. The main result is that a weak connectednes property (c.2)) combined with a strong continuity property (the product topology finer than the order topology) is equivalent to a strong connectedness property (c.3) combined with a weak continuity property (called continuity).

In section 4.4 it is shown how these properties imply the existence of real continuous order homomorphisms. The important results hold only for at most countable products.

4.2 Independence assumptions

The totally preordered set $Y = \prod_{i \in I} Y_i$, where I is an index set, is given.

Definition 65 (mixing) *The mixing mapping*

$$\boxtimes : Y \times Y \times 2^I \to Y$$

is defined by

$$y = y' \boxtimes_A y''$$
$$\text{where}$$
$$y_i = \begin{cases} y'_i \text{ for } i \in A \\ y''_i \text{ for } i \in A^c \end{cases}$$

Remark 18 $\boxtimes$ *will trivially have the propety that*

$$y' \boxtimes_A y'' = y'' \boxtimes_{A^c} y'$$

Denote for $A \subset I, \prod_{i \in A} Y_i$ by $Y_A, \prod_{i \notin A} Y_i$ by Y_{A^c}, and by $(y_A, y'_{A^c}) = y \boxtimes_A y', (y_A \in Y_A, y'_{A^c} \in Y_{A^c},)$

The totally ordered quotient set $(Y/\sim, \succeq)$ is denoted $(S, \succeq)$.

For $A = \{i\}$ we shall denote an element in $Y_{A^c} = Z_i = Y_{\{i\}^c}$ by z_i (and obviously $y_i \in Y_i$) so $y \boxtimes_i y = y = (y_i, z_i) \in Y$.

Definition 66 **($\succsim_{A,w}$)** *For a given $w \in Y_{A^c}, Y_A$ is a totally preordered set with the preorder $\succsim_{A,w}$defined by*

$$\begin{gathered} y_A \succsim_{A,w} y'_A \text{ if} \\ y \boxtimes_A w \succsim y' \boxtimes_A w \end{gathered}$$

Definition 67 (independence with respect to A) *$(Y, \succsim)$ is independent with respect to A if $\succsim_{A,w}$on Y_A is independent of w.*

Definition 68 (independence with respect to $\mathcal{A}$) *$(Y, \succsim)$ is independent with respect to $\mathcal{A}$ if $(Y, \succsim)$ is independent with respect to A for all $A \in \mathcal{A}$.*

In particular for $A = \{i\}$ and $\mathcal{A} = (\{i\})$.

Definition 69 (factor independence) *The preorder $\succsim$ on Y is factor independent if for all $i \in I$, $\succsim_{z_i}$on Y_i is independent of $z_i \in Z_i$.*

$(Y_i, \succsim_{z_i})$ will under the factor independence assumption be denoted $(Y_i, \succsim)$. $\prod Y_i$ will have the product preorder (see definition 29). The identity mapping

$$1 : \prod Y_i \to Y$$

will be a homomorphism

$$\prod (Y_i, \succsim) \to (Y, \succsim)$$

In later chapters the independence assumptions will turn out to have important consequences for the existence of real additive representations. In this chapter only the weakest assumption - factor independence - will be studied.

4.3 Order topologies on product sets

$(Y, \precsim)$ and $(S, \succeq)$ will have order topologies (section 2.4 page 20). $(Y_i, \precsim_{z_i})$ will also have an order topology for all $z_i \in Z_i$. Factor independence will of course imply that these topologies are independent of z_i. The sections $Y_i \times \{z_i\}$ will be subsets of Y and therefore have a relative topology. The topology induced on Y_i by the projection $(y_i, z_i) \longmapsto y_i$ will also be called the relative topology (on Y_i for the given z_i). Y is the product $\prod_{i \in I} Y_i$, and will have a product topology whenever a topology is chosen on $(Y_i)_{i \in I}$.

Definition 70 (the product topology) *Assuming factor independence we get* **the** *product topology on Y, when we chose the order topology on $(Y_i, \precsim)$.*

Definition 71 (c.1) $(Y, \precsim)$ *will have this property if Y and S are connected in their order topologies.*

Definition 72 (c.2) $(Y, \precsim)$ *will have this property if $(Y_i, \precsim_{z_i})$ for all $i \in I$ and all $z_i \in Z_i$ are connected in their order topologies.*

Remark 19 *By lemma 2 page 22 c.1 (c.2) holds if and only if $(Y, \precsim)$ and $(S, \succeq)$, $((Y_i, \precsim_{z_i}))$ are order-complete and have no gaps.*

Definition 73 (c.3) $(Y, \precsim)$ *will have this property if $(Y_i, \precsim_{z_i})$ for all $i \in I$ and all $z_i \in Z_i$ are connected in the relative topology.*

The relative topology on Y_i is finer than the order topology (for any $z_i \in Z_i$) so c.3 implies c.2. It is easy to check (see chapter 2) that c.3 will hold if and only if for all $i \in I$ and all $z_i \in Z_i$, Y_i is order-complete, has no gaps, and $(y_i'', z_i) \precsim (y_i', z_i') \precsim (y_i''', z_i)$ imply that there exists a $y_i \in Y_i$ such that $(y_i, z_i) \sim (y_i', z_i')$. c.3 also implies that the relative topology on Y_i is equal to the order topology for all $z_i \in Z_i$. It will be a consequence of theorem 8 page 34 that some assumptions on I and Y will imply that c.3 implies c.1. The following examples all with $I = \{1, 2\}$ show that we cannot get any stronger theorem relating c.1, c.2, and c.3.

Example 5 *Let $Y_1 = [0,1] \cup [2,3]$ and $Y_2 = [0,1]$; $\precsim$ is defined by $(y_1', y_2') \precsim (y_1'', y_2'')$ if*

$$y_1' + y_2' \geqq y_1'' + y_2''$$

c.1 holds but not c.2 and c.3.

Example 6 *Let* $Y_1 = [0,1]$ *and* $Y_2 = [0,1]$; $\succsim$ *is defined by* $(y_1', y_2') \succsim (y_1'', y_2'')$ *if*

$$\begin{aligned} y_1' + y_2' \geqq y_1'' + y_2'' \text{ for } y_1' + y_2' \text{ or } y_1'' + y_2'' \neq 1 \\ y_1' \geqq y_1'' \text{ for } y_1' + y_2' = y_1'' + y_2'' = 1 \end{aligned}$$

Now c.1 and c.2 hold but not c.3

Example 7 *Let* $Y_1 =]0,1[$ *and* $Y_2 =]0,1[$; $\succsim$ *is defined by* $(y_1', y_2') \succsim (y_1'', y_2'')$ *if*

$$\begin{aligned} y_1' + y_2' \geqq y_1'' + y_2'' \text{ for } y_1' + y_2' \text{ or } y_1'' + y_2'' \neq 1 \\ y_1' \geqq y_1'' \text{ for } y_1' + y_2' = y_1'' + y_2'' = 1 \end{aligned}$$

Now c.2 holds but not c.1 or c.3.

Theorem 8 *Let* $(Y, \succsim) = \left(\prod_{i \in I} Y_i, \succsim\right)$ *be a totally preordered product set, where* I *is a finite set, and assume factor independence. Then* $(Y, \succsim)$ *has property c.3 if and only if it has property c.2 and the order topology is coarser than the product topology.*

Proof. $\{y \in Y \,|\, y \succ y_0\}$ is by c.3 equal to (with y_i'' the first element in Y_i)

$$\bigcup_{y' \sim y \sim y''} \{y \in Y \,|\, y_i \succ y_i' \text{ or } y_i \succsim y_i'' \text{ for all } i \in I\}$$

and therefore open. This and the dual gives the only if part of the theorem.

If the product topology is finer than the order topology, the sets

$$\{y_i \in Y_i \,|\, (y_i, z_i) \succ y'\} \text{ and } \{y_i \in Y_i \,|\, (y_i, z_i) \prec y'\}$$

are open in the order topology on Y_i and c.2 implies that one of the sets is empty or there exists a $y_i \in Y_i$ such that $y' \sim (y_i, z_i)$ and $(Y, \succsim)$ has property c.3 ■

In order to generalize theorem 8 we need an extra property.

Definition 74 ($\succsim$ continuous) $\succsim$ *on* $Y = \prod_{i \in I} Y_i$ *is continuous if for all* $y, y' \in Y, y \boxtimes_{A_n} y' \to y$ *for* $I \backslash A_n \to \emptyset$.

Y will have this continuity property if the product topology is finer than the order topology.

Lemma 7 *Let $Y = \prod_{i \in I} Y_i$ be a product set where I is countable. Let $\succsim$ be a total preorder on Y, and assume factor independence and continuity. If for all $i \in I, y_i \succsim y'_i$, then $y \succsim y'$. If for all $i \in I, y_i \succsim y'_i$ and for some $i \in I, y_i \succ y'_i$, then $y \succ y'$.*

Proof. Denote by $\mathbf{n}$ the set $\{1, 2, \cdots, n\} \subset I$.

$$y \succsim y' \boxtimes_{\mathbf{n}} y \succsim y' \boxtimes_{\mathbf{n+1}} y \to y'$$

so $y \succsim y'$ and analogously ■

Lemma 8 *Let $Y = \prod_{i \in I} Y_i$ be a product set where I is countable. Let $\succsim$ be a total preorder on Y, and assume factor independence. Then continuity of $\succsim$ implies that the convergence $y' \boxtimes_{\mathbf{n}} y \to y'$ is uniform in $y \in Y$.*

Proof. Assume that the convergence $(y'_{\mathbf{n}}, y_{\mathbf{n}^c}) \to y'$ is not uniform in y. This means that there exists sequences $y^k \in Y, n_k, (k = 1, 2, \cdots, n_k \to \infty)$ and an open interval $\mathcal{O}$ containing a $y' \in Y$ such that $y' \boxtimes_{\mathbf{n}_k} y^k \notin \mathcal{O}$. Define

$$y''_i \text{ (and } y'''_i\text{) by } y''_i = \max_{\mathbf{n}_k \leqq i} y^k_i \left(y'''_i = \min_{\mathbf{n}_k \leqq i} y^k_i \right)$$

Then by lemma 7

$$y' \boxtimes_{\mathbf{n}_k} y'' \succsim y' \boxtimes_{\mathbf{n}_k} y^k \succsim y' \boxtimes_{\mathbf{n}_k} y'''$$

for all k, but $y' \boxtimes_{\mathbf{n}_k} y''$ and $y' \boxtimes_{\mathbf{n}_k} y'''$ converges to y' and we have a contradiction ■

Theorem 9 *Let $(Y, \succsim) = (\prod_{i \in I} Y_i, \succsim)$ be at totally preordered product set, where I is a countable set, and assume factor independence. Then $(Y, \succsim)$ is continuous and has property c.3 if and only if it has property c.2 and the product topology is finer than the order topology.*

Proof. If a sequence converges in the product topology it also converges in a coarser topology, so it is trivial that $(Y, \succsim)$ is continuous when the product topology is finer than the order topology. c.3 follows as in the proof of theorem 8.

To prove that c.3 and continuity imply that the product topology is finer than the order topology we will prove that a set of the form

$\{y \in Y \,|y \succ y_0\}$ is open in the product topology. Let $y \succ y_0$ and $y \succ y_1 \succ y_0$, the set $\prod_{i=1}^{n}]y_i', \rightarrow[\times \prod_{i=n+1}^{\infty} Y_i$, where y_i' is chosen such that $y_i' \prec y_i$ and $(y_{\mathbf{n}}', y_{\mathbf{n}^c}) \succ y_1$, will be open in the product topology, will contain y, and will by lemma 8 for n sufficiently large be contained in $\{y \in Y \,|y \succ y_0\}$ ■

The connectedness property and factor independence have the surprising consequence that Y_i is separable.

Lemma 9 *Factor independence and c.3 implies that $(Y_i, \precsim)$ is separable in the order topology. (Assuming that $Y_j / \sim$ has more than one element for some $j \in I \setminus \{i\}$).*

Proof. Chose some $w \in Y \,|\{i,j\}^c$ .Let $y_0, y_1 \in Y_j$ $y_1 \succ y_0$ and choose sequences $x_1, x_2, \cdots$ and $x_0, x_{-1}, \cdots$ such that $(x_i, y_0) \sim (x_{i-1}, y_1) \in Y_i \times Y_j$. These sequences are finite if $(x_n, y_1) \succ (x, y_0)$ (or $(x_{-n}, y_0) \prec (x, y_1)$) for some n and for all x. We want now to prove that if for example $x_1, x_2, \cdots$ is infinite then for all x there exists an N such that $x_i \succ x$ for $i > N$. If not then $x_i \uparrow x'$ and $(x_i, y_0) \sim (x_{i-1}, y_1) \in s_i \in (Y_i \times Y_j) / \sim$. Theorem 8 then implies that $(x', y_0) \sim (x', y_1)$, and this contradicts factor independence. If we can find a countable dense set in $[x_0, x_1]$ (or any other interval $[x_i, x_{i+1}]$) and possibly $[x_n, \rightarrow[$ and $]\leftarrow, x_{-n}]$ we shall know that Y_i is separable. Define $x_{0.5}$ and $y_{0.5}$ by $(x_{0.5}, y_{0.5}) \sim (x_0, y_1) \sim (x_1, y_0)$ and $(x_{0.5}, y_0) \sim (x_0, y_{0.5})$. Using the same technique we can define $x_{0.25}, x_{0.75}$ etc. We get in this way a countable subset of $[x_0, x_1]$. Assume that $]a, b[\subset [x_0, x_1]$ does not contain any element from this set. Then there exists sequences (x_i') , (y_i') and (x_i'') with $x_i' \precsim a \prec b \precsim x_i''$, with (x_i') increasing and (y_i') and (x_i'') decreasing, such that either $(x_i', y_i') \sim (x_{i+1}'', y_{i+1}') \sim (x_i'', y_0)$, $x_{i+1}' = x_i'$ and $(x_i', y_{i+1}') \sim x_{i+1}'', y_0$ or $(x_i', y_i') \sim (x_{i+1}', y_{i+1}') \sim (x_i'', y_0)$, $x_{i+1}'' = x_i''$ and $(x_i', y_{i+1}') \sim (x_{i+1}', y_0)$. Then $x_i' \uparrow x_0'$, $y_i' \downarrow y_0'$ and $x_i'' \downarrow x_0''$. Using theorem 8 we get $(x_0', y_0') \sim (x_0', y_0') \sim (x_0'', y_0)$, and there is a violation of factor independence. The set $\{x_{0.5}, x_{0.25}, x_{0.75}, \cdots\}$ is therefore dense in $[x_0, x_1]$. A similar technique can be used for the first and last intervals. For for example the last interval we can replace y_1 by $y_1' = \sup \{y \,|(x, y_0) \sim (x_n, y) \, , x \in Y_i\}$ ■

4.4 Existence of real continuous order homomorphisms

As an application of theorem 5 page 23 and corollary 2 we shall now prove

Theorem 10 *Let $Y = \prod_{i \in I} Y_i$ be a totally preordered product set where at least two of the factors have non equivalent elements. Assume*

1) $\succsim$ *is factor independent*
2) *c.3*
3) *Either I is finite, or I is countable and $\succsim$ continuous*

Then there exists real order homomorphisms

$$g : Y \to [0,1] \text{ and } g_i : Y_i \to [0,1], i \in I$$

Proof. Assume that Y and $(Y_i)_{i \in I}$ have first and last elements. Lemma 9 says that Y_i is separable and theorem 9 that Y_i is connected in the order topology. Theorem 5 then for each $i \in I$ gives an order isomorphism f_i between $Y_i/\sim$ and the closed unit interval. c.3 and for I countable continuity of $\succsim$ implies (theorems 9 and 8) that the product topology is finer than the order topology so Y and thus $Y/\sim$ are therefore also connected and separable in the order topology. There exists then an order isomorphism and homeomorphism f on $Y/\sim$ with values in the unit interval. Denote by χ ($\chi_i, i \in I$) the canonical mapping $Y \to Y/\sim$ $(Y_i \to Y_i/\sim)$. Then $g : Y \to [0,1]$ $(g_i : Y_i \to [0,1])$ defined by $g = f \circ \chi$ $(g_i = f_i \circ \chi_i)$ are the order homomorphisms. If Y and/or Y_i have no first or last element the same proof with the obvious modification applies ∎

Corollary 3 *The function $h : [0,1]^I \to [0,1]$ defined by*

$$h\left((g_i(y_i))_{i \in I}\right) = g\left((y_i)_{i \in I}\right)$$

is continuous and strictly monotonic.

Proof. Trivial ∎

4.5 Note

Theorem 10 is apparently new. Wold (1943-44) [188] Part II axiom V.2 page 223 and Wold (1952) [189] assumption B, page 82, imply c.3. These representation theorems are therefore special cases of theorem 10.

5

A subset of a product set

5.1 Introduction

This chapter begins the study of the simplest possible structure on a product set.

I is a finite index set, $(S_i)_{i \in I}$ are arbitrary sets, and

$$Q \subset S = \prod_{i \in I} S_i$$

An important application of this structure is to the case where Q is the graph of a relation on a product set.

The axiom which makes this an interesting structure is introduced in section 5.2. It is the independence condition on Q. Some regularity conditions are also introduced. In section 5.3 the result that this implies the existence of an independent total preorder on $S = \prod_{i \in I} S_i$, such that $Q = \{s \in Q \,|\, s \succ s_0\}$ for some $s_0 \in S$, is formulated. In chapter 9 theorem 36 page 113 this theorem is proved and extended to the surprising main theorem in this book. It is shown that there exists a bijection f between $S = \prod_{i \in I} S_i$ and (a subset of) $\mathbb{R}^I$, such that $f(Q) = \left\{x \in f(S) \,\middle|\, \sum_{i \in I} x(i) > 0\right\}$. Applying this central result to the case, where Q is the graph of a relation, gives representation theorems in the form of additive real functions for order relations on function spaces, both for total and for non-total relations. Two conditions known to imply additivity of representations in the case of a total preorder on $S_1 \times S_2$, the Thomsen and the Reidemeister conditions, are in section 5.4 page 43 shown to be special cases of the independence conditions.

5.2 Independence

Notation 3 *The notation $s_A \in S_A$ will be used for elements s_A in the product $\prod_{i \in A} S_i = S_A$ for $A \subset I$.*

Definition 75 (essential) *$s_A \in S_A$ is essential for Q if there exist $s = s \boxtimes_A s \in Q$ and $s \boxtimes_A s' \notin Q$. In particular $s_i \in S_i$ is essential if there exists $(s_j)_{j \neq i}, (s'_j)_{j \neq i}$ such that $s \boxtimes_i s = s \in Q$ and $s \boxtimes_i s' \notin Q$. Q is essential if for all $i \in I$ and all $s_i \in S_i$, s_i is essential.*

Remark 20 *Q essential obviously does not imply that all s_A are essential.*

Definition 76 (independence) *Q is independent with respect to $A \subset I$ if for all $s_A, s'_A \in S_A$ and all $s_{A^c}, s'_{A^c} \in S_{A^c}$*

$$\begin{array}{cc} s = s \boxtimes_A s \in Q & s \boxtimes_A s' \notin Q \\ s' \boxtimes_A s \notin Q & s' = s' \boxtimes_A s' \in Q \end{array} \tag{NOT}$$

can not hold.

Definition 77 (independence) *Q is independent with respect to $\mathcal{A} \subset 2^I$ or just independent if it is independent with respect to all $A \in \mathcal{A}$.*

Q independent with respect to $A \subset I$ is trivially equivalent to Q independent with respect to A^c.

5.3 A total preorder on the set S_A

Definition 78 ($\succ_A$ on S_A) *The relation $\succ_A$ is defined on the essential elements from S_A by*

$$s_A \succ_A s'_A \text{ if there exist } s_{A^c} \text{ with } (s_A, s_{A^c}) \in Q \text{ and } (s'_A, s_{A^c}) \notin Q$$

Definition 79 ($\sim_A$ on S_A) *The relation $\sim_A$ is defined on the essential elements from S_A by*

$$s_A \sim_A s'_A \text{ if } (s_A, s_{A^c}) \in Q \text{ if } (s'_A, s_{A^c}) \in Q$$

Remark 21 *$\sim_A$ is clearly an equivalence relation.*

Definition 80 **($\succsim_A$ on S_A)** *The relation $\succsim_A$ is defined on the essential elements from S_A by*

$$\succsim_A = \succ_A \cup \sim_A$$

Lemma 10 *Let Q be independent with respect to A. Then $\succsim_A$ is a total preorder on the essential elements from S_A. $(S_A, \succsim_A)$ is independent (with respect to all subsets of A).*

Proof. $\succsim_A$ is reflexive because $\sim_A$ is. If $s_A \succsim_A s'_A \succsim_A s''_A$ one can by going through the cases prove that $s_A \succsim_A s''_A$, so $\succsim_A$ is transitive. Let for example $s_A \succ_A s'_A \succ_A s''_A$, then there exists s_{A^c}, s'_{A^c}, such that

$$\begin{array}{ll} s \boxtimes_A s \in Q & s'' \boxtimes_A s' \notin Q \\ s' \boxtimes_A s \notin Q & s' \boxtimes_A s' \in Q \end{array}$$

then by the independence assumption $s \boxtimes_A s' \in Q$ so $s_A \succ s''_A$. It is clear from the definition that the relation is total. Either $s_A \succ_A s'_A$, $s_A \prec_A s'_A$, or $s_A \sim_A s'_A$ will hold for any essential pair $s_A, s'_A \in S_A$. $\succ$ is asymmetric, this is simply another way of formulating the independence condition

The independence of $(S_A, \succsim_A)$ with respect to all subsets of A follows directly from the independence assumption ■

Q implies thus - under the independence condition - the existence of order relations on the factors of the product S. The factors have therfore order topologies and the set S itself **the** product topology.

Lemma 11 *Let S and Q be open and S_i connected, then for all essential s_A there exists an s_{A^c} and an open neighborhood $\mathcal{N}$ of s such that all elements from $\mathcal{N}$ restricted to any proper subset $B \subset I$ are essential.*

Proof. For any $i \in I$ chose $s_{\{i\}^c}$ on the boundary of Q - this is possible because of c.3 - then for a sufficiently small neighborhood of s_i there will be an open neighborhood $\mathcal{N}_i$ where $s_{\{i\}^c}$ will be essential. This implies s_B essential in $\bigcup_{i \notin B} proj \mathcal{N}_i$, where the projection is on S_B ■

Remark 22 *The relation $\succsim_A$ may be a total preorder even if Q is not independent with respect to A.*

Definition 81 (Q factor independent) *If $\succsim_{\{i\}}$ is a total preorder for all $i \in I$, Q is called factor independent.*

The consequences of the independence assumptions depend on $\#I$.

$\#I = 1$. No elements are essential, and the independence condition is empty.

$\#I = 2$. The meaning of essential is obvious, and the independence conditions is the factor independence condition from chapter 4 page 32. It means that the two sets have total preorders, and that the subset has the monotonicity property that $s \in Q$ ($s \notin Q$) implies $s' \in Q$ ($s' \notin Q$) for $s' \precsim s, (s \precsim s')$. The consequences of factor independence were studied in chapter 4.

$\#I = 3$. This case is much like $\#I = 2$. There are no reasons to expect that there is a preorder on S such that the preorders on S_A for $A \subset I$ can be derived from this preorder as the restriction to sections. It is easy to construct examples where no such preorder exist.

$\#I > 3$. In this case the independence condition with respect to 2^I on (S, Q) becomes a very strong condition with important consequences. The first step behind the characterization of independent relations on a function space is this theorem.

Theorem 11 (equivalence) *Let*

$$Q \subset S = \prod_{i \in I} S_i$$

where $\#I > 3$, $\#S_i > 1$ for all $i \in I$.
The following two statement are equivalent
(a) Q is independent, essential and open and S_i connected
(b) $(S_A, \succsim_A)_{A \in \mathcal{A}}$ are order-complete, totally preordered spaces with no gaps, derived from an independent total preorder $\succsim$ on S and $Q = \{s \in Q \,|\, s \succ s_0\}$ for some $s_0 \in S$.

Proof. See chapter 9[1] ■

Remark 23 *This theorem makes it possible to translate all results about additivity representations of total preorders to results about additivity results for relations which are not total.*

---

[1]It is an open problem if there is a direct proof of this theorem. The proof given in chapter 9 uses the representation theorems.

5.4 The Thomsen and the Reidemeister conditions

In this section it is assumed that S is a product $S_1 \times S_2 \times T_1 \times T_2$, and that $S_1 \times S_2 = T_1 \times T_2$ so Q is the graph of a relation on $S_1 \times S_2$, the relation will be denoted $\mathcal{P}$. The factor independence assumption for Q and the topological assumptions from theorem 11(a) are made in this section.

The diagonal of $S_1 \times S_2 \times T_1 \times T_2$ will be denoted Δ

$$\Delta = \{(s_1, s_2, t_1, t_2) \in S_1 \times S_2 \times T_1 \times T_2 \,|(s_1, s_2) = (t_1, t_2)\}$$

For most or all applications of the results based on the independence assumptions we can assume

Definition 82 (A) *Assumption A is*

$$\Delta \subset \overline{Q} \setminus Q = \overline{graph\mathcal{P}} \setminus graph\mathcal{P} \tag{A}$$

The assumption says that

$$s \in \overline{\mathcal{P}(s)} \text{ but } s \notin \mathcal{P}(s) \qquad \text{(local non-satiation)}$$

The assumption will therefore also be made here even if the results hold with obvious modifications without this assumption.

Denote the coordinates in $S_1 \times S_2 \times T_1 \times T_2$ by $(1, 2, 1', 2') = X$. The independence conditions with respect to two-elements subsets of X now becomes

I.(1,2)

$$\begin{matrix} (s,t) \in Q & (s',t) \notin Q \\ (s,t') \notin Q & (s',t') \in Q \end{matrix} \qquad \text{(NOT (1,2))}$$

This means according to lemma 10 that there are total preorders $\succsim_S$ and $\succsim_T$ on S and T such that $(s, t) \in Q$ implies $(s', t') \in Q$ for $s' \succsim_S s$ and $t' \succsim_T t$. If assumption A holds $(S, \succsim_S) = (T, \precsim_T)$ and Q is the graph of the $\succ_S$ relation (denoted $\succ$) on S, so $(s,t) \in Q$ is equivalent to $s \succ t$.

I.(1,1')

$$\begin{matrix} (s,t) \in Q & (s_1', s_2, t_1', t_2) \notin Q \\ (s_1, s_2', t_1, t_2') \notin Q & (s',t') \in Q \end{matrix} \qquad \text{(NOT (1,1'))}$$

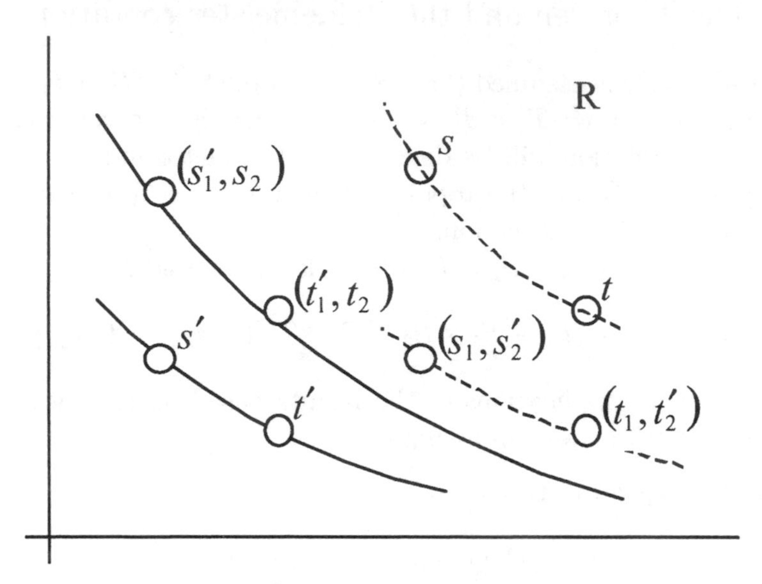

FIGURE 5.1. Reidemeister

Definition 83 (*R*) *Conditions NOT (1,2) and NOT (1,1') together is called the Reidemeister condition or just R.*

R means that

$$\left.\begin{array}{r} s \succsim t \\ s' \succsim t' \\ (t_1, t_2') \succsim (s_1, s_2') \end{array}\right\} \Rightarrow (s_1', s_2) \succsim (t_1', t_2) \tag{R}$$

I.(1,2')

$$\begin{array}{cc} (s, t) \in Q & (s_1', s_2, t_1, t_2') \notin Q \\ (s_1, s_2', t_1', t_2) \notin Q & (s', t') \in Q \end{array} \tag{NOT (1,2'))}$$

Definition 84 (generalized Thomsen) *Conditions NOT (1,2) and NOT (1,2') together is called the Generalized Thomsen condition or just GT.*

GT means that

$$\left.\begin{array}{r} s \succsim t \\ s' \succsim t' \\ (t_1', t_2) \succsim (s_1, s_2') \end{array}\right\} \Rightarrow (s_1', s_2) \succsim (t_1, t_2') \tag{GT}$$

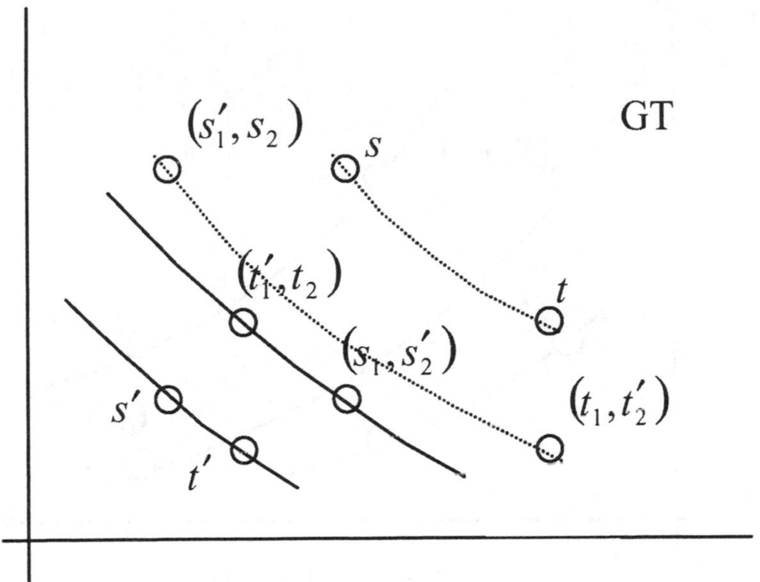

FIGURE 5.2. Generalized Thomsen

Definition 85 (Thomsen) *The special case*

$$(t_1', t_2) = (s_1, s_2')$$

yields the original Thomsen condition

$$\left.\begin{array}{c} s \succsim t \\ s' \succsim t' \end{array}\right\} \Rightarrow (s_1', s_2) \succsim (t_1, t_2') \tag{T}$$

See figure 5.3 page 46

Definition 86 (Briancon) *As a special case of both T and R we obtain the Briancon condition or just B*

$$\left.\begin{array}{c} s \succsim t \succsim u \\ (t_1, u_2) \succsim (s_1, t_2) \end{array}\right\} \Rightarrow (t_1, s_2) \succsim (u_1, t_2) \tag{B}$$

5.5 Note

5.5.1 The Reidemeister and Thomsen conditions

The conditions T, R, and B were introduced by W. Blaschke, G. Thomsen, and K. Reidemeister around 1928, in connections with

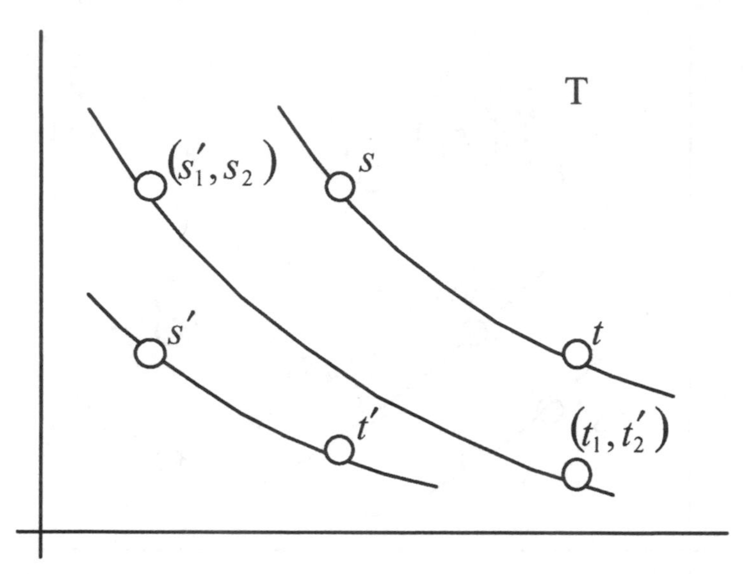

FIGURE 5.3. Thomsen

discussions of nets[2] see Reidemeister (1929) [155], Thomsen (1927 and 1929) [173] and [174]. A 3-net consists of three sets of curves in the plane such that two curves have at most one point in common and such that one and only one curve from each family goes through each point. If it is possible to find a homeomorphism such that the three families become sets of parallel straight lines T, R, and B will obviously hold. Part of the converse result was proved in Thomsen (1927) [173] under differentiability assumptions. The converse result that B implies 7.1 is more general than theorem 20 because T implies B, on the other hand $X \times Y$ is not - in our results - assumed to be a subset of $\mathbb{R}^2$. Blaschke (1934) [29] and Blaschke Bol (1938) [30] give further information on the theory of nets.

The observation that T, GT, R, and B are independence conditions and special cases of independence conditions for the graph of a relation on a product of two sets comes from [180].

[2] Also called webs (German Gewebe)

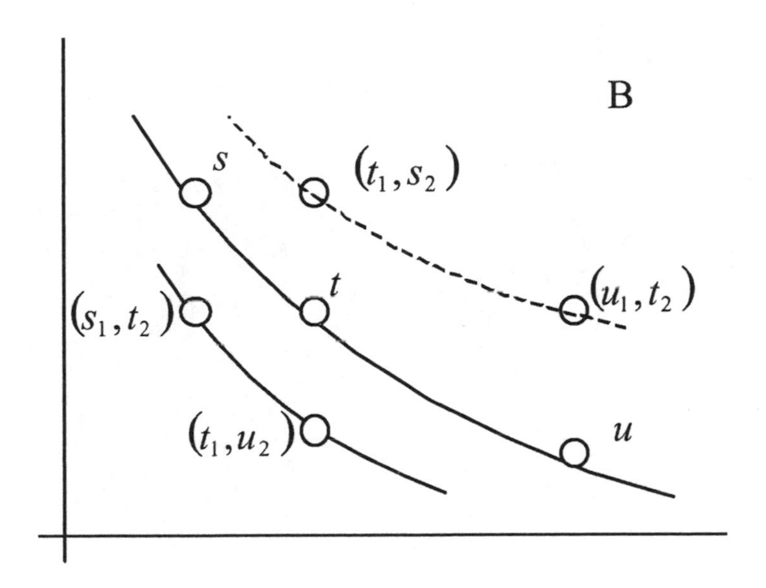

FIGURE 5.4. Briancón

6

Mean groupoids

6.1 Introduction

The main problem in chapter 2 was the study of a set S with a total preorder $\succsim$. This chapter studies a set S with an order relation $\succeq$ and an algebraic operation $\circ$ on S. Intuitively $a \circ b$ can be thought of as the mean of or the midpoint between a and b, $a \circ a = a$, and if $a \succ b$, then $a \succ a \circ b \succ b$. Later chapters will be concerned with totally preordered sets that have enough structure to define a mean operation on $((S/\sim), \succeq)$.

The properties of $\circ$ that make $(S, \succeq)$ a commutative mean groupoid will be given section 6.2. In section 6.3 page 52 it is proved that $(S, \succeq, \circ)$ can be completed. Section 6.4 page 54 contains the main theorem of this chapter - the Aczél Fuchs theorem - theorem 13 page 55. It shows that any commutative mean groupoid is isomorphic with a subspace of $(\mathbb{R}, \geqq, \circ)$ where $r_1 \circ r_2 = \frac{1}{2}(r_1 + r_2)$. Theorem 14 in section 6.5 shows that if $(S, \succeq)$ is connected, and $\circ$ is defined locally on S, and has the mean groupoid properties where defined, then $\circ$ can be uniquely extended to all of S making $(S, \succeq, \circ)$ a commutative mean groupoid. Section 6.6 shows how the results in sections 6.4 and 6.5 are related to the bisymmetry equation from the theory of functional equations.

6.2 Definition of a commutative mean groupoid

$(S, \succeq)$ is a totally ordered set and $\circ$ is binary operation also defined on S.

Definition 87 (commutative mean groupoid) *$(S, \succeq, \circ)$ is a commutative mean groupoid if for all $a, b, c, d \in S$.*

A.1	$a \circ b$	$\in$	S	
A.2	$a \circ a$	$=$	a	*idempotent*
A.3	$a \circ b$	$=$	$b \circ a$	*commutative*
A.4	$(a \circ b) \circ (c \circ d)$	$=$	$(a \circ c) \circ (b \circ d)$	*bisymmetry*

and

M.1	$a \succ b$	$\Rightarrow$	$a \circ c \succ b \circ c$ *and* $c \circ a \succ c \circ b$
M.2	$a \succ b \succ c$	$\Rightarrow$	$\exists n \in \mathbb{Z}_+$, *such that* $na \circ c \succ b \succ a \circ nc$
where	$na \circ c$	$=$	$\underbrace{a \circ (a \circ \cdots \circ (a \circ c))}_{n \text{ times}}$ *and analogously for* $a \circ nc$

A.1 says that the mean of two elements in S is again an element in S.

A.2 means that every element in S is idempotent, i.e. the mean of a and a is a.

A.3 says that $\circ$ is commutative, i.e. the mean of a and b is equal to the mean of b and a. (If all axioms except A.3 hold S is defined to be a mean groupoid).

A.4 is a bisymmetry[1] property. It says that the mean of the mean of $(a \circ b)$ and $(c \circ d)$ is equal to the mean of $(a \circ c)$ and $(b \circ d)$. This property is the most difficult to check in the applications later in this book.

M.1 says that if $a \succ b$, then the mean of a and any other element c will be greater than the mean of b and c. This is a monotonicity assumption.

M.2 is an Archimedean property. If $a \succ b \succ c$, then by forming the mean of a and the mean of a and $\cdots$ and the mean of a and c one eventually gets an element $na \circ c \succ b$, and dually.

Definition 88 (isomorphism) *A strictly monotonic bijection*

$$f : (S, \succeq, \circ) \to (S', \succeq, \circ)$$

[1]Bisymmetry has also been called medial, entropic, and even alternation. Medial appears to be the most common teminology, but I have kept the Aczél teminology from mean groupoids.

is an isomorphism if

$$a \circ b = c \Leftrightarrow f(a) \circ f(b) = f(c)$$

Definition 89 (isomorphic) *Two commutative mean groupoids* $(S, \succeq, \circ)$ *and* $(S', \succeq, \circ)$ *are isomorphic if there exists an isomorphism between them.*

It is easy to see that $a \succ b$ implies that $a \succ a \circ b \succ b$ (M.1). This means that $(S, \succeq)$ has no gaps (see page 11).

It is also easy to see that

$$a \circ (b \circ c) = (a \circ b) \circ (a \circ c) \quad \text{(A.4)}$$

and that

$$(a \circ b) = (a \circ c) \Longrightarrow b = c \text{ (M.1).}$$

Example 8 *Let* S *be an interval of* $\mathbb{R}$*, let* $\succeq$ *be* $\geqq$*, and define* $\circ$ *by*

$$r_1 \circ r_2 = \frac{1}{2}(r_1 + r_2)$$

It is obvious that $(S, \geqq, \circ)$ *is a commutative mean groupoid. If* $\mathbb{R}$ *is replaced by* $\mathbb{Q}$ *we shall still get the same result. In general any subset* $\mathbb{R}_1 \subset \mathbb{R}$ *for which* $r_1, r_2 \in \mathbb{R}_1$ *implies* $\frac{1}{2}(r_1 + r_2) \in \mathbb{R}_1$ *is a commutative mean groupoid. Theorem 13 (page 55) will show that this is the most general commutative mean groupoid in the sense that any commutative mean groupoid* $(S, \succeq, \circ)$ *will be isomorphic to* $(\mathbb{R}_1, \geqq, \circ)$ *for some such* $\mathbb{R}_1 \subset \mathbb{R}$.

The following lemma gives a useful relation between M.2 and continuity.

Lemma 12 *Let* $(S, \succeq, \circ)$ *be a totally ordered set with the order topology and a binary operation* $\circ$ *which has properties A.1,2,4, and M.1. Define*

$$F : S \times S \to S \text{ by } F(a, b) = a \circ b$$

Then

$$M.2 \Rightarrow F \text{ is continuous}$$
$$S \text{ order-complete and } F \text{ continuous} \Rightarrow M.2$$

Proof. Assume M.2. F is strictly monotonic in both variables (M.1) so it is enough for a given $b \in S$ to prove for example continuity of the strictly monotonic g defined by $g(a) = F(a,b)$. $g^{-1}(c \prec c_0)$ is either open or equal to $\{a \,|a \preceq a_0\}$ for some a_0. If $g^{-1}(c \prec c_0) = \{a \,|a \preceq a_0\}$,we have $a_0 \circ b \prec c_0$ and $a \circ b \succ c_0$ for all $a \succ a_0$. For a given $a \succ a_0$,.$n(a_0 \circ b) \circ (a \circ b) = a_n \circ b \succeq c_0$ for all n, because $a_n = na \circ c \succ a_0$, so we have a contradiction with M.1 or M.2 and $g^{-1}(c \prec c_0)$ is open. $g^{-1}(c \succ c_0)$ is by duality also open. Since sets of the form $g^{-1}(c \prec c_0)$ and $g^{-1}(c \succ c_0)$ form a subbase for the order-topology, g is continuous.

Assume now that S is order-complete[2] and F continuous. Let $a \succ b \succ c$ and define $a_n = na \circ c$, then $a_n \uparrow a_0$ for some $a_0 \preceq a$. Assume $a \succ a_0$, continuity of F implies $a_{n+1} = F(a_n, a) \uparrow F(a_0, a) = a_0 \circ a \succ a_0$, this is a contradiction and $a_0 = a$. This means that $a_n \succ b$ for some $n > N$ for some N. The other part follows analogously ■

6.3 Completion of commutative mean groupoid

It is shown in chapter 2 subsection 2.2.2 that a totally ordered set $(S', \succeq)$ can be completed. It is the purpose of this section to show that if S' is a commutative mean groupoid $(S', \succeq, \circ)$, the completion $(S, \succeq)$ - with a natural definition of $\circ$ on S - is also a commutative mean groupoid.

Definition 90 **($\circ$ on S)** *$\circ$ on S is now defined as*

$$a \circ b = c$$

where

$$a = (S_a, S^a), b = (S_b, S^b)$$

and c is the cut determined by

$$(S_a \circ S_b, S^a \circ S^b)$$

See definition 32 page 15

$$(S_a \circ S_b = \{s \in S' \,|a \circ b = s, \ a \in S_a, \ b \in S_b\})$$

and analogously

---

[2] Definition 18 page 12

Lemma 13 *A.1, A.2, A.4, M.1 and M.2 for* $(S', \succeq, \circ)$ *imply that* $a \circ b$ *is a cut in* S'*, so A.1 holds for* S

Proof.

$$S_a \circ S_b \preceq S^a \circ S^b$$

follows from M.1, so the only problem is to check that there does not exist $c_1 \prec c_2$ such that

$$S_a \circ S_b \preceq c_1 \prec c_2 \preceq S^a \circ S^b$$

Assume that there exists $c_1 \prec c_2$ such that

$$a' \circ b' \preceq c_1 \prec c_2 \preceq a'' \circ b''$$

for all

$$a' \in S_a, a'' \in S^a, b' \in S_b, b'' \in S^b$$

Then - using A.2, M.1 and A.4 -

$$\begin{gathered} a' \circ b' \preceq c_1 \circ (a' \circ b') \prec c_2 \circ (a' \circ b') \preceq \\ (a'' \circ b'') \circ (a' \circ b') = (a'' \circ a') \circ (b'' \circ b') \end{gathered}$$

This holds for fixed a', b', and for all a'', b''. M.2 implies that there exists a'', b'' such that $(a'' \circ a') \in S_a$ and $(b'' \circ b') \in S_b$. Then

$$c_2 \circ (a' \circ b') \preceq (a'' \circ a') \circ (b'' \circ b') = a_1' \circ b_1' \preceq c_1$$

by induction

$$\begin{gathered} nc_2 \circ (a' \circ b') \preceq \left(a''_{n-1} \circ b''_{n-1}\right) \circ \left(a'_{n-1} \circ b'_{n-1}\right) = \\ \left(a''_{n-1} \circ a'_{n-1}\right) \circ \left(b''_{n-1} \circ b'_{n-1}\right) = a_n \circ b_n \preceq c_1 \end{gathered}$$

for all n, thus violating M.2 ■

Theorem 12 (completion) *If* $(S', \succeq, \circ)$ *is a commutative mean groupoid, the completion* $(S, \succeq, \circ)$ *is then an order-complete commutative mean groupoid.*

Proof. A.2 holds for S', so

$$S_a \circ S_a = S_a \text{ and } S^a \circ S^a = S^a$$

this means that A.2 holds for S.

A.3 holds for S', so

$$S_a \circ S_b = S_b \circ S_a \text{ and } S^a \circ S^b = S^b \circ S^a$$

this means that A.3 holds for S.

A.4 holds for S', so

$$(S_a \circ S_b) \circ (S_c \circ S_d) = (S_a \circ S_c) \circ (S_b \circ S_d)$$

and

$$\left(S^a \circ S^b\right) \circ \left(S^c \circ S^d\right) = (S^a \circ S^c) \circ \left(S^b \circ S^d\right)$$

Using M.2 and the same technique as in the proof of lemma 13 on

$$(S_a \circ S_b) \circ (S_c \circ S_d), \left(S^a \circ S^b\right) \circ \left(S^c \circ S^d\right)$$

and

$$(S_a \circ S_c) \circ (S_b \circ S_d), (S^a \circ S^c) \circ \left(S^b \circ S^d\right)$$

implies that these pairs determines cuts and A.4 holds for S.

M.2 for S' trivially implies M.2 for S.

M.1 Finally let A.1, A.2, A.3, M.1, and M.2 hold for S', $a \succ b$ implies that $S_a \supset S_b$ and that $S_a \backslash S_b$ contains at least two elements. $a \circ c$, $[b \circ c]$ is the cut determined by

$$((S_a \circ S_c), (S^a \circ S^c)), \left[\left((S_b \circ S_c), \left(S^b \circ S^c\right)\right)\right]$$

and M.2 for S' implies

$$S_a \circ S_c \supset S_b \circ S_c$$

and that $(S_a \circ S_c) \backslash (S_b \circ S_c)$ contains at least two elements so $a \circ c \succ b \circ c$ for S, and M.1 holds for S ∎

6.4 The Aczél Fuchs theorem

We are now able to state and prove the main theorem of this chapter. In other parts of the book this theorem will be applied to cases where a totally preordered space $(S, \succsim)$ has enough extra properties to define a mean groupoid operation on $(S/\sim, \succeq)$.

Let $\mathbb{R}_1$ denote a subset of $\mathbb{R}$ with the property that $\frac{1}{2}(r_1 + r_2) = r_1 \circ r_2 \in \mathbb{R}_1$ for $r_1, r_2 \in \mathbb{R}_1$.

Theorem 13 (Aczél Fuchs)

$$(S, \succeq, \circ) \textit{ is a commutative mean groupoid}$$
$$\Leftrightarrow$$
$$(S, \succeq, \circ) \textit{ and } (\mathbb{R}_1, \geqq, \circ) \textit{ are isomorphic}$$

Proof. The $\Leftarrow$ part of theorem is obvious.

Assume first that $(S, \succeq)$ is order-complete. We construct the isomorphism f first on a subset $S' \subset S$, then on the closure of S', and finally on the rest of S.

Two arbitrary points $a, b \in S$ are selected with $a \prec b$. Define using the binary number system $s_0 = a, s_1 = b$, and $S' = \{s_0, s_1, s_{0.1}, s_{0.01}, \cdots\}$ where $s_{0.1} = s_0 \circ s_1, s_{0.01} = s_0 \circ s_{0.1}$ and in general

$$s_\xi = s_{\xi+2^{-n}} \circ s_{\xi-2^{-n}}$$

where $\xi 2^n$ is an odd integer.

$$(\xi = 0.\xi_1\xi_2 \ldots \xi_n 000 \ldots, \xi_i \in \{0, 1\}, i = 1, 2, \ldots, n, \xi_n = 1)$$

Next define

$$f : (S', \succeq, \circ) \to \mathbb{R} \text{ by } f(s_\xi) = \xi$$

It has to be proved that $f : S' \to f(S')$ is a strictly monotonic bijection with

$$f(a \circ b) = \frac{1}{2}(f(a) + f(b)) \qquad (\maltese)$$

f is obviously strictly monotonic.

Let $\xi = 0.\xi_1\xi_2 \ldots \xi_n 000 \ldots, \xi_n = 1$ and $\eta = 0.\eta_1\eta_2 \ldots \eta_n 000 \ldots$, or $\eta = 1$, then assuming $(\maltese)$ hold for $n - 1$

$$\begin{aligned}
s_\xi \circ s_\eta &= \\
\left(s_{\xi+2^{-n}} \circ s_{\xi-2^{-n}}\right) \circ \left(s_{\eta+\eta_n 2^{-n}} \circ s_{\eta-\eta_n 2^{-n}}\right) &= \quad \text{(Definition)} \\
\left(s_{\xi+2^{-n}} \circ s_{\eta-\eta_n 2^{-n}}\right) \circ \left(s_{\xi-2^{-n}} \circ s_{\eta+\eta_n 2^{-n}}\right) &= \quad \text{(A.3 and A.4)} \\
s_{\frac{1}{2}(\xi+\eta)+2^{-(n-1)}(1-\eta_n)} \circ s_{\frac{1}{2}(\xi+\eta)-2^{-(n-1)}(1-\eta_n)} &= \quad \text{(Induction hyp.)} \\
s_{\frac{1}{2}(\xi+\eta)}
\end{aligned}$$

$(\maltese)$ is thus true for n if is true for $n - 1$ and for $n = 1$. For $n = 1, s_{0.1} = s_0 \circ s_1$, so $(\maltese)$ holds for S'. f is strictly monotonic and $f(S')$ is dense in $[0,1]$ so a cut in S' determines a cut in $[0,1]$. We define $f(s)$, where s is a cut in S' as the corresponding cut in $[0,1]$. In order to prove that all elements in $[a,b]$ are cuts in S' and that f is strictly monotonic let

$$a \succ c \succ d \succ b, (c, d \notin S')$$

and assume that no $s \in S'$ with $c \succ s \succ d$ exists. M.2 gives the existence of an n such that

$$c \prec a \circ nd$$

and then for all $x \in S', x \prec c$ and all $y \in S', y \succ d$ we shall have

$$x \prec a \circ ny$$

and from (✠) which holds for S'

$$f(x) < f(a \circ ny) = (1 - 2^{-n}) f(y)$$
$$(\{f(x) | x \in S', x \prec c\}, \{f(y) | y \in S', y \succ d\})$$

is a cut in $f(S')$ which is dense in $[0,1]$, so we have a contradiction, and f is defined and strictly monotonic on $[a,b]$. Denote $f^{-1}(x)$ by s_x for $x \in [0,1]$.

To prove (✠) on S we note that S' is dense in $[a,b]$, that $a \circ b$ determines a continuous function, and that (✠) holds on S', it will therefore also hold on $[a,b]$. For $c \prec a$ and $b \circ c \succeq a$, we define $f(c) = -f(c')$ where c' is the element in $[a,b]$ such that $c \circ c' = a$.

For $c \succ b$ (or $c \prec b', b' \circ b = a$) we get from $A.2.$ that there exists an n such that

$$a \prec c' = na \circ c \prec b \,(\text{or } a \succ c' = na \circ c \succ b')$$

and we define $f(c) = 2^n f(c')$. It is easy to verify that this definition is independent of n and that the extension is strictly monotonic. (✠) follows from

$$f(c \circ d) = 2^n f(c' \circ d') =$$
$$2^{n-1} (f(c') + f(d')) = \tfrac{1}{2} (f(c) + f(d))$$

If $(S, \succeq, \circ)$ is not order-complete a function f can be constructed on the completion of S. The restriction to S will have all the required properties ■

Remark 24 *If $g(a) = \alpha + \beta f(a), \alpha, \beta \in \mathbb{R}, \beta > 0$, then g will be another isomorphism i.e. strictly monotonic function for which (✠) holds. It is clear from the construction of f that no other such functions will exist. f is determined uniquely as soon as $f(s_0)$ and $f(s_1)$ have been chosen.*

For some applications there is in a mean groupoid a natural zero denoted $\Box$.

Definition 91 *$(S, \succeq, \circ, \Box)$ where $\Box \in S$ is a commutative mean groupoid with a zero.*

Remark 25 *In a representation of a commutative mean groupoid with a zero $\Box$ it will be natural to chose $f(\Box) = 0$ and then of course the α in the previous remark 24 can only be chosen to be 0.*

Remark 26 *f is an order isomorphism and therefore a homeomorphism in the order topologies. (See lemma 3 page 22).*

S has no gaps and if it is order-complete the range $f(S)$ of the homeomorphism f is an interval.

If $(S', \succeq, \circ)$ is not order-complete $f(S')$ will be dense in $f(S)$, where S is the completion of S'.

If S has a first (last) element $f(S)$ will have a first (last) element. If S have no first (last) element $f(S)$ may or may not have a minorant (majorant).

6.5 Extension of a commutative mean groupoid

In applications of the mean operation $\circ$ it is not always natural or possible to assume that $a \circ b$ is defined for all $a, b \in S$. It is, however, easy to prove the following theorem.

Theorem 14 *Let $(S, \succeq, \circ)$ be a connected totally ordered set where for any $a \in S$, $b \circ c$ is defined for $b, c \in I_a$, where I_a is an open interval containing a. If $\circ$ on I_a has the properties A.2-A.4 and M.1 and M.2, then it can be uniquely extended to all of S. $(S, \succeq, \circ)$ with the extended $\circ$ will be a connected commutative mean groupoid.*

Proof. Let $a \in S$ and I_a an open interval containing a in which $b \circ c$ is defined. $(I_a, \succeq, \circ)$ is then a commutative mean groupoid, and there exists (theorem 13 page 55) an f_a determined by its values for two points in I_a. Let $c \in S, [a, c]$ or $[c, a]$ is covered by open sets $I_b, b \in [a, c] \cup [c, a]$. Since S is connected, S is order-complete (lemma 2 page 22) and $[a, c] \cup [c, a]$ is compact (theorem 4 page 21). So $[a, c] \cup [c, a] \subset \bigcup_{i=1}^{n} I_{b_i}$ where $I_{b_i} \cap I_{b_{i+1}}$ contains at least two points. With $a = b_1$ and f_a chosen, $f_{b_{i+1}}$is determined by the value of f_{b_i}in the two points in $I_{b_i} \cap I_{b_{i+1}}$ for $i = 1, 2, \cdots n$. Assume that f_1 and f_2 are two functions chosen in this way with the same $(I_a)_{a \in S}$ but different I_{b_i}. The restriction of f_1 and f_2 to I_a will be identical. Assume now that d is the infimum (supremum) of the elements larger (smaller) than a for which an open neighborhood contains elements for which f_1 and f_2 are different. There will exist an open interval I_d containing d on which f_1 and f_2 are isomorphisms for the same commutative mean groupoid. This is clearly a contradiction. We can finally define

$$a \circ b = f^{-1} \left(\frac{1}{2} \left(f\left(a\right) + f\left(b\right) \right) \right)$$

The extended $\circ$ will according to the "if" part of theorem 13 page 55 be a commutative mean groupoid ■

As a step in the proof we have proved a combination of theorems 13 and 14. Let $(S, \succeq, \circ)$ be a connected totally ordered set where for any $a \in S, b \circ c$ is defined for $b, c \in I_a$, where I_a is an open interval containing a. We state this result as

Theorem 15 *Let $(S, \succeq, \circ)$ be a connected totally ordered set, where for any $a \in S, b \circ c$ is defined for $b, c \in I_a$, where I_a is an open interval containing a. If $(S, \succeq, \circ)$ has properties A.2, A.3, A.4, M.1, and M.2 where $\circ$ is defined, then there exists a real order preserving function f such that*

$$f\left(a \circ b\right) = \frac{1}{2} \left(f\left(a\right) + f\left(b\right) \right) \qquad (**)$$

whenever $a \circ b$ is defined.

Proof. Take f as the restriction of the f from the proof of theorem 14. If (**) holds for the extended $\circ$ it will also hold for $\circ$ ■

6.6 The bisymmetry equation

A "translation" of the results in the previous sections give some results from the theory of functional equations.

Given a totally ordered set $(S, \succeq)$. The problem is to characterize the set of continuous, strictly increasing functions

$$F : S \times S \to S$$
for which
$$F(a,a) = a, F(a,b) = F(b,a) \tag{6.1}$$
and
$$F(F(a,b), F(c,d)) = F(F(a,c), F(b,d))$$

The last equation in 6.1 is called the bisymmetry equation and is well-known as a substitute for associativity in some algebraic theories.

Define $\circ$ on S by $a \circ b = F(a,b)$.

According to lemma 12 page 51, when S is order-complete, continuity of F is equivalent to M.2; F strictly increasing is equivalent to M.1 etc. This means that $(S, \succeq, \circ)$ is a commutative mean groupoid.

In the terminology of functional equations theorem 13 page 55 can be rewritten as follows:

Theorem 16 *Let $(S, \succeq)$ be a connected totally ordered set. The set of F's for which*

$$F(a,b) = f^{-1}\left(\frac{1}{2}(f(a) + f(b))\right) \tag{6.2}$$

for some continuous, strictly monotonic real function f is equal the set of continuous, strictly monotonic F's for which $F(a,a) = a, F(a,b) = F(b,a)$ and $F(F(a,b), F(c,d)) = F(F(a,c), F(b,d))$

Proof. Trivial ■

Continuous, strictly monotonic real functions f's are simpler than F, and the characterization 6.2 is therefore called the solution of the functional equation 6.1.

Theorem 17 *Let $(S, \succeq)$ be a connected totally ordered set and F a strictly monotonic function*

$$F : \bigcup_{c \in S} I_c \times I_c \to S$$

where I_c is an open interval containing c. Assume that F where defined has the properties

$$F(a,a) = a, F(a,b) = F(b,a)$$

and

$$F(F(a,b), F(c,d)) = F(F(a,c), F(b,d))$$

Then F can - retaining its properties - be extended to $S \times S$

Proof. Trivial given theorem 14 ■

6.7 Notes

6.7.1 History and other results

This chapter is a small example of the general theory of sets with both an order structure and an algebraic structure see Fuchs (1963) [82]. Hölder (1901) [98] proved the main classical result in this field. His theorems give a condition (similar to M.2) on totally ordered commutative groups for the existence of an isomorphism to a subgroup of the real numbers (as an additive group with the natural order). It has later turned out that the restriction to *commutative* groups is unnecessary.

Axiom A.4 has been used in Aczél (1948) [3]. He stated and proved the Aczél Fuchs theorem for the case where S is a subset of the real numbers. Fuchs (1950) [81] showed that S can be any totally ordered set.

Fuchs has proved that the only consequence of dropping A.3 for the Aczél Fuchs theorem is that (✠) becomes

$$f(a \circ b) = \lambda f(a) + (1-\lambda) f(b), \lambda \in [0,1]$$

Aczél has shown that dropping A.3 and A.2 would give

$$f(a \circ b) = \lambda f(a) + \mu f(b) + \gamma, (\lambda, \mu > 0)$$

instead of (✠).

The formulation of all these results in terms of functional equations can be found in Aczél (1966) [5].

6.7.2 Classifying commutative mean groupoids

Theorem 13 shows that there are ten different order-complete commutative mean groupoids isomorphic with

$[0]$	$]-\infty,\infty[$	$]-\infty,0[$	$]-\infty,1]$	$[0,\infty[$
$[0,1[$	$[0,1]$	$]0,\infty[$	$]0,1[$	$]0,1]$

respectively. If the commutative mean groupoids are not order-complete, they will be isomorphic with a dense subspace of one of the nine non-degenerate intervals, but they are more difficult to characterize using a Hamel basis than totally ordered Archimedean groups.

6.7.3 Lexicographic "mean groupoids"

If S is the unit square with the lexicographic order and $\circ$ is defined by $(x,y)\circ(x',y')=\frac{1}{2}(x+x',y+y')$, then $(S,\geqq,\circ)$ will have all the properties of a commutative mean groupoid except M.2. In general if $(S_\alpha,\geqq_\alpha,\circ_\alpha)_{\alpha<\mu}$ are sets of commutative mean groupoids for all ordinals $\alpha<\mu$,where μ is an ordinal, then $\prod_{\alpha<\mu}S_\alpha$ with the lexicographic order will have the properties of a commutative mean groupoid except M.2. $\prod_{\alpha<\mu}S_\alpha$ will be isomorphic with $\prod_{\alpha<\mu}\mathbb{R}_\alpha$ where $\mathbb{R}_\alpha$ is a dense subspace of one of the nine non-degenerate intervals. If all totally ordered sets $(S,\succeq,\circ)$ for which A.1-4 and M.1 hold are isomorphic with $\prod_{\alpha<\mu}S_\alpha$ for some μ and some choice of $\mathbb{R}_\alpha$, we should have a theorem for "non-Archimedean commutative mean groupoids" analogous to the classical Hahn theorem for totally ordered groups. See Hahn (1907) [92] and Fuchs (1963) [82].

6.7.4 Totally ordered mixture spaces

von Neumann Morgenstern (1943) [141] (with a proof of existence in the second edition 1947) base their "measurement of utility" on a structure similar to the mean groupoid structure. In view of the historical importance of their results for the development of results in chapter 10 we shall - without details - indicate the precise relation between the two structures. A *totally ordered mixture space* is a totally ordered set with an algebraic operation for each $\lambda\in[0,1]$. With the axioms of von Neumann and Morgenstern a totally ordered mixture space with the algebraic operation only for $\lambda = \frac{1}{2}$ is isomorphic with a commutative mean groupoid. On the other hand a commutative mean groupoid is isomorphic with a dense subspace of a totally ordered mixture space with $\circ$ mapped into the algebraic operation for $\lambda = \frac{1}{2}$. The operation for dyadic numbers can be defined by induction and then extended by continuity to all $\lambda \in [0, 1]$. The conclusion is that a totally ordered mixture space contains unnecessarily much structure for the purpose in mind. It is enough to have the algebraic operation defined for $\lambda = \frac{1}{2}$. See also Herstein Milnor (1953) [95] and Cramér (1956) [45].

6.7.5 Reducible

An $F : S \times S \rightarrow S$ is defined to be reducible if $b \neq c$ implies $F(a, b) \neq F(a, c)$ and $F(b, a) \neq F(c, a)$. It is easy to see that strict monotonicity of a continuous F under the assumption that S is connected implies reducibility. If F is reducible and continuous, it will be strictly monotonic or strictly decreasing ($b \succ c$ implies $F(a, b) > F(a, c)$ for all a or $F(a, b) < F(a, c)$ and analogously). This equivalence means that the strict monotonicity assumption in theorem 16 can be replaced by the reducibility assumption.

6.7.6 Products of mean groupoids

We shall see in theorem 19 page 74 that products of mean groupoids may be of interest.

Definition 92 *Let $(S, \succeq, \circ) = \prod_{i \in I} (S_i, \succeq_i, \circ_i)$ be a product of totally ordered spaces (with a midpoint operation defined on S_i) with an order relation $\succeq$ defined by $\succeq_i$ for all $i \in I$ and a midpoint $\circ$ defined by $s' \circ s'' = s$ for $s'_i \circ s''_i = s_i$ for all $i \in I$*

The resulting space has all the properties of commutative mean groupoids except that the order is not total. There may be some hope that conversely any "partially ordered mean groupoid" is isomorphic with a product of mean groupoids. For corresponding results for partially ordered groups see Lorenzen (1950) [129].

6.7.7 Completion

The completion result - theorem 12 - in this chapter may be new. The proof was simplified by Erik Christensen, who objected to the earlier version. The generalization to mean groupoids (without axiom A.3) is trivial.

6.7.8 Measurement of magnitudes

Theorems giving isomorphisms or homomorphisms between a set S and a subset of $\mathbb{R}$ with part of the usual structure are usually presented as the basis of *measurement of magnitudes.* S and the structure on S usually contains a total (pre)order(a larger (longer, heavier, louder, warmer) than b).

Addition is usually considered as another part of the structure. This leads either to the assumption that S is unbounded, and this is unnatural for some interpretations, or to results which can not easily be expressed as isomorphism theorems. (See for example Bourbaki (1939 -) [33] General Topology, chapter V § 2, propositions 1 and 2). In view of this, and of the results later in this book, the Aczél Fuchs theorem appears to give a basis for "measurement" which should be preferred for sets without gaps.

6.7.9 The bisymmetry equation

The bisymmetry equation (or A4)

$$F\left(F\left(a,b\right),F\left(c,d\right)\right)=F\left(F\left(a,c\right),F\left(b,d\right)\right) \qquad \text{(bisymmetry)}$$

is the essential axiom for the Aczél Fuchs theorem. It has been suggested by Samuelson (1992) [159] that it could be replaced by a more intuitive condition for a midpoint operation, namely the functional equation

$$f\left(x,y\right)=f\left(f\left(x,f\left(x,y\right)\right),f\left(f\left(x,y\right),y\right)\right) \qquad (6.3)$$

or that A.4 could be replaced with A.4'

$$a\circ b=\left(a\circ\left(a\circ b\right)\right)\circ\left(\left(a\circ b\right)\circ b\right) \qquad \text{(A.4')}$$

Some attempts to prove or find counter examples to the proposition that 6.3, under some extra assumption, implies

$$f\left(x,y\right)=g^{-1}\left(\frac{1}{2}\left(g\left(x\right)+g\left(y\right)\right)\right) \qquad (6.4)$$

for some $g : X \to \mathbb{R}$ are summarized below.

Let X be an arbitrary set and $f : X \times X \to X$. Obvious solutions to 6.3 are then (for $h : X \to X$ and $g : X \to \mathbb{R}$)

$$\begin{array}{ll} f(x,y) \in h^{-1}(h(x)) \text{ where } h(h(x)) = x & \text{(a)} \\ f(x,y) \in h^{-1}(h(y)) \text{ where } h(h(y)) = y & \text{(b)} \\ f(x,y) \in g^{-1}(-g(x) - g(y)) & \text{(c)} \\ f(x,y) \in g^{-1}\left(\frac{1}{2}(g(x) + g(y))\right) & \text{(d)} \end{array}$$

If h or g are injective, $\in$ will be replaced by $=$ and $h(x) = x$. In (c) and (d) $\mathbb{R}$ could be replaced by an Abelian group G.

If there is a total order $\succeq$ on X we get the solutions

$$\begin{array}{ll} f(x,y) = \max(x,y) & \text{(e)} \\ f(x,y) = \min(x,y) & \text{(f)} \\ f(x,y) = \begin{cases} f_1(x,y) & \text{for } x \succ y \\ g(x) & \text{for } x = y \\ f_2(x,y) & \text{for } x \prec y \end{cases} & \text{(g)} \end{array}$$

where g is any function for which

$$g(x) = f(f(x, g(x)), f(g(x), x))$$

and f_1 (and f_2) is any solution to 6.3 for which

$$x \succ f_1(x,y) \succ y \, (y \succ f_2(x,y) \succ x)$$

If f is assumed to be increasing in both variables it is easily seen that (6.3) implies $f(x,x) = x$ and that most of the solutions are excluded. If f is strictly increasing only (6.4) and (g) are still solutions. If the condition $f(x,y) = f(y,x)$ is added only (6.4) is left as a solution.

If X is a lexicographically ordered product, we may get (g) for each coordinate, but there is no hope for (g) as a solution, so some continuity or Archimedean assumption is needed for (g) or (6.4)

We can now formulate the following

Conjecture 1 *Let $(X, \succeq)$ be a connected totally ordered set. Let $f : X \times X \to X$ be a strictly increasing continuous function for which (6.3) holds. Then (6.4) holds for some (strictly increasing and continuous) functions $g : X \to \mathbb{R}$. (The topology on X is the order topology).*

The continuity and standard technique imply that we, without loss of generality, can regard X as a subset of $\mathbb{R}$ and get (6.4) to hold with $g(x) = x$ for points of the form

$$\left(\frac{2p \pm 1}{2^{n+1}}, \frac{2p \mp 1}{2^{n+1}}\right) \text{ and } \left(\frac{p \pm 1}{2^n}, \frac{p \mp 1}{2^n}\right)$$

where p is an odd integer and $n \in \mathbb{Z}$. If (g) or (6.4) holds in a uniform neighborhood of the diagonal in $X \times X$ it holds everywhere. The possible deviations from (g) depends on X not being bounded.

One of the main arguments (note 6.7.8) for using $(\mathbb{R}, \geqq, \circ)$ and sets isomorphic with subsets of this set as a foundation for "measurement theory" instead of $(\mathbb{R}, \geqq, +)$ (as in [33] General Topology pages 12-17) is that we do not have to assume that X is unbounded. So if (6.3) implies (6.7.9) or (6.4) only under unboundedness assumptions this would be a further argument against (6.3).

References to attempts to find functional equations with the solution (6.4) can be found in [5] footnote 95 page 279.

A more general problem

$$\left(M(x_1, x_2, \cdots, x_n) = f\left(\sum f^{-1}(x_i)\right)\right)$$

has been treated in an unpublished dissertation by I. Fenyö in 1945. He makes however strong extra assumptions, so if these assumptions are needed (6.3) is still no reasonable substitute for (6.1).

6.7.10 Counter example (Andrew Gleason, Harvard)[3]

For $\lambda > 0, x \in \mathbb{R}$ define

$$\varphi(\lambda, x) = x + \frac{1}{2k} \sin(k \log \lambda) \sin x$$

where

$$k = \frac{2\pi}{\log 2} \simeq 9.065$$

Note that

$$\varphi\left(\frac{1}{2}\lambda, x\right) = \varphi(\lambda, x) = \varphi(2\lambda, x)$$

[3] Letter of May, 1983 to Paul Samuelson, copy to me summer 2000.

$\varphi(\lambda, \cdot) \in C^{\infty}$ and invertible, since

$$D_2\varphi(\lambda, x) = 1 + \frac{1}{2k}\sin(k \log \lambda)\cos x \geq \frac{1}{2}$$

so the inverse function is also C^{∞}.

Theorem 18 (Gleason) *If $x < y$, there is a unique value of λ such that*

$$\varphi(\lambda, x) + \lambda = \varphi(\lambda, y)$$

Proof. Omitted ■

Now define

$$f(x, y)$$

by

$$\varphi(\lambda_0, f(x, y)) = \frac{1}{2}\lambda_0 + \varphi(\lambda_0, x)$$

using the unique λ from theorem 18.

To verify (6.3), note that to calculate $f(x, f(x, y))$ we start by solving the equation

$$\varphi(\lambda, x) + \lambda = \varphi(\lambda, f(x, y))$$

for λ. Because of the periodicity of φ in λ the solution is

$$\lambda = \frac{1}{2}\lambda_0$$

and so $f(x, f(x, y))$ is determined by

$$\begin{aligned} \varphi\left(\tfrac{1}{2}\lambda_0, f(x, f(x, y))\right) &= \varphi\left(\tfrac{1}{2}\lambda_0, x\right) + \tfrac{1}{4}\lambda_0 \\ \varphi(\lambda_0, f(x, f(x, y))) &= \varphi(\lambda_0, x) + \tfrac{1}{4}\lambda_0 \end{aligned}$$

Similarly, we see that

$$\varphi(\lambda_0, f(f(x, y), x)) = \varphi(\lambda_0, x) + \frac{3}{4}\lambda_0$$

This shows that to evaluate

$$f(f(x, f(x, y)), f(f(x, y), y)$$

we again choose $\lambda = \frac{1}{2}\lambda_0$ and we get

$$\varphi\left(\lambda_0, f\left(f\left(x, f\left(x, y\right)\right), f\left(f\left(x, y\right), y\right)\right)\right) = \\ \varphi\left(\lambda_0, x\right) + \tfrac{1}{2}\lambda_0 = \varphi\left(\lambda_0, f\left(x, y\right)\right)$$

and (6.3) is established.

$x = 0, y = \frac{1}{2}$, and $z = \frac{3}{4}$ after some computations gives $f\left(0, \frac{3}{4}\right) \neq \frac{3}{8}$, so we have a counter example to Samuelson's conjecture.

A Theorem

In an E-mail of July 5, 2000 Gleason informs me that he now has a proof of Samuelson's conjecture assuming that $f \in C^2$.

Conclusion

Samuelson's conclusion (see [159]) is that this shows that we have to assume differentiability. My conclusion is that we have to assume bisymmetry, because that means that we do not have to assume differentiability.

7

Products of two sets as a mean groupoid

7.1 Introduction

The problem in this chapter is to study a totally preordered product set $(X \times Y, \succsim)$[1]. In chapter 4 theorem 10 page 37 the existence of a real representation is proved.

In section 7.4 conditions will be given for the existence of real order homomorphisms $f : X \times Y \to \mathbb{R}$, $f_1 : X \to \mathbb{R}$, $f_2 : Y \to \mathbb{R}$ such that

$$f(x, y) = f_1(x) + f_2(y) \tag{7.1}$$

It is first shown that $S = X \times Y / \sim$ under the independence condition on the graph of $\succsim$ is a commutative mean groupoid. The independence conditions are in chapter 5 shown to be that the relation is a total preorder and to yield the Thomsen condition and the Reidemeister condition. The results from chapter 6 then imply that (7.1) holds.

If a real function F is given on a set $X \times Y$, we can define $\succsim$ on $X \times Y$ by

$$(x, y) \succsim (x', y') \text{ if } F(x, y) \geqq F(x', y')$$

The result obtained from the Aczél-Fuchs theorem can therefore be used to obtain conditions for the existence of a solution to the functional equation

$$F(x, y) = g^{-1}(f_1(x) + f_2(y)) \tag{7.2}$$

---

[1] To avoid trivial exceptions it is assumed that X and Y have more than one element each.

Real functions of two variables have of course many applications, and in many of these applications will it be of great interest to know if (7.1) or (7.2) hold.

The main importance of the results are however that they can be used to find additive representation of relations on function spaces. See chapters 8 and 9.

7.2 Thomsen's and Reidemeister's conditions

The representation of $(X \times Y, \succsim)$ from theorem 10 is of course not always of the form (7.1). (7.1) implies that the equivalence classes in $(X \times Y, \succsim)$ have certain properties, and it is obvious that $(X \times Y, \succsim)$ will not have these properties in general.

In this section we shall discuss some of these conditions for (7.1) and in section 7.4 prove that one of these conditions is in fact equivalent to (7.1). It is proved by defining a midpoint operation $\circ$ on $(S, \succeq)$ and showing hat this makes $(S, \succeq, \circ)$ a commutative mean groupoid. (Section 7.3).

$(X \times Y, \succsim)$ is in this section a totally preordered set for which factor independence and c.3 holds.

If (7.1) holds it is obvious that

$$\begin{array}{ll} (x_0, y_1) \sim (x_1, y_0) & \text{and} \\ (x_0, y_2) \sim (x_2, y_0) & \text{imply} \\ (x_1, y_2) \sim (x_2, y_1) & \end{array} \qquad \text{(Thomsen)}$$

$$\left(\begin{array}{ll} f_1(x_0) + f_2(y_1) = f_1(x_1) + f_2(y_0) & \text{and} \\ f_1(x_0) + f_2(y_2) = f_1(x_2) + f_2(y_0) & \text{imply} \\ f_1(x_1) + f_2(y_2) = f_1(x_2) + f_2(y_1) & \end{array} \right)$$

This is the Thomsen[2] condition known from chapter 5. It is the independence condition on the graph of $\succsim$ with respect to $(1, 2'), (1', 2)$, where $(1, 2, 1', 2')$ denote factors in the product

$$(X \times Y \times X \times Y).$$

[2]See note 5.5.1 for an explanation of the name Thomsen.

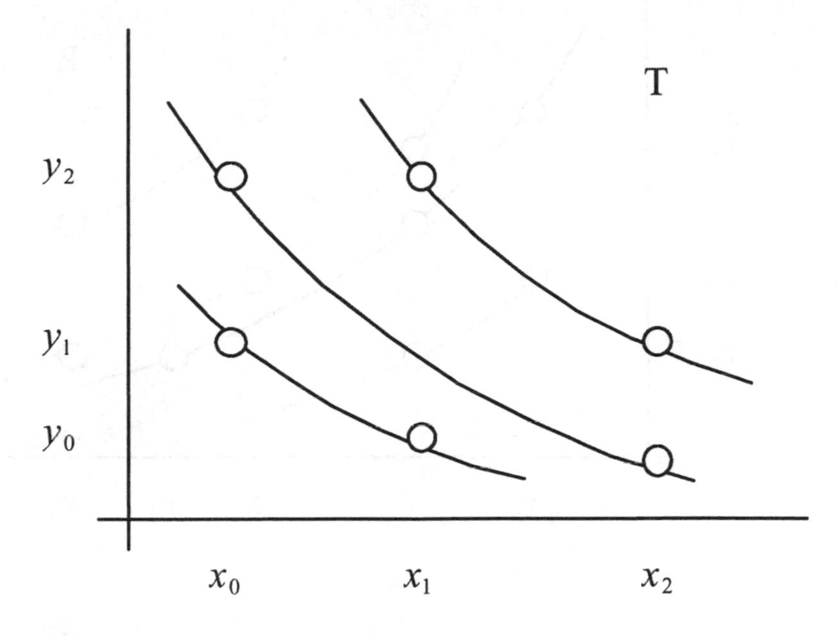

FIGURE 7.1. Thomsen

It is equally obvious that (7.1) implies that

$$\begin{array}{ll} (x_1, y_2) \sim (x_2, y_1), & \\ (x_1, y_4) \sim (x_2, y_3) & \text{and} \\ (x_3, y_2) \sim (x_4, y_1) & \text{imply} \\ (x_3, y_4) \sim (x_4, y_3) & \end{array} \qquad \text{(Reidemeister)}$$

$$\left(\begin{array}{ll} f_1(x_1) + f_2(y_2) = f_1(x_2) + f_2(y_1), & \\ f_1(x_1) + f_2(y_4) = f_1(x_2) + f_2(y_3) & \text{and} \\ f_1(x_3) + f_2(y_2) = f_1(x_4) + f_2(y_1) & \text{imply} \\ f_1(x_3) + f_2(y_4) = f_1(x_4) + f_2(y_3) & \end{array} \right)$$

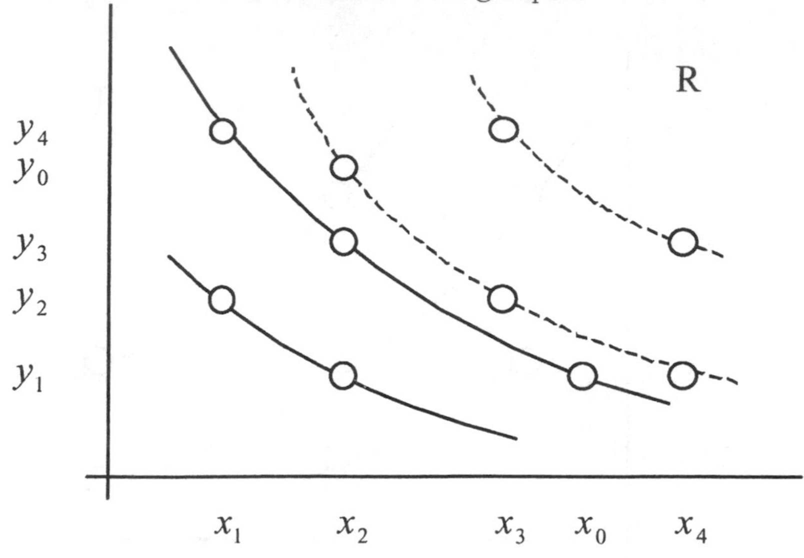

FIGURE 7.2. Reidemeister

This is the Reidemeister[3] condition known from chapter 5. It is the independence condition for the graph of $\succsim$ with respect to $((1,1'),(2,2'))$.

We recall that the independence of a graph of a relation with respect to $((1,2),(1',2'))$ means that the relation is a total preorder (or two total preorders, which coincide if the diagonal in $X \times Y \times X \times Y$ is on the boundary of the graph). See chapter 5.

Lemma 14 *T implies R.*

Proof. Let $x_1 \prec x_2 \prec x_4, x_1 \prec x_3 \prec x_4, y_1 \prec y_2 \prec y_4, y_1 \prec y_3 \prec y_4$. Other cases can be covered by duality and a change in subscripts. T and R are symmetric in X and Y so we can assume $(x_1, y_4) \precsim (x_4, y_1)$. There will then exist an x_0 such that $x_2 \prec x_0 \precsim x_4$ and $(x_0, y_1) \sim (x_1, y_4) \sim (x_2, y_3)$.

[3]See note 5.5.1 for an explanation of the name Reidemeister.

1. Assume that there exists a y_0 such that

$$\begin{array}{llll} \text{I} & (x_1, y_2) \sim & (x_2, y_1) & \\ \text{II} & (x_1, y_4) \sim & (x_2, y_3) & \sim (x_0, y_1) \\ \text{III} & (x_2, y_0) \sim & (x_3, y_2) & \sim (x_4, y_1) \end{array}$$

T gives then applied to

$$\begin{array}{lll} \text{I and II} & & (x_0, y_2) \sim (x_2, y_4) \\ \text{(I and II)} & \text{and III} & (x_0, y_0) \sim (x_3, y_4) \\ \text{II and III} & & (x_4, y_3) \sim (x_0, y_0) \end{array}$$

so we get $(x_4, y_3) \sim (x_3, y_4)$ and R is proved for this case.

2. If $(x_2, y) \prec (x_3, y_2)$ for all y we have to construct sequences of elements $x_1^i, x_2^i \in X$ such that for $(i = 1, 2, \cdots)$

$$x_1^1 = x_1, x_2^1 = x_2 \textit{ and } (x_1^i, y_2) \sim (x_1^{i-1}, y_4) \sim (x_2^i, y_1) \sim (x_2^{i-1}, y_3)$$

1. gives

$$(x_1^i, y_4) \sim (x_2^i, y_3)$$

Factor independence implies as in the proof of lemma 9 that

$$(x_1^n, y_4) \sim (x_2^n, y_3) \succ (x_3', y_2) \sim (x_4', y_1)$$

where

$$(x_3', y_4) \sim (x_4', y_3) \sim (x_3, y_2) \sim (x_4, y_1)$$

for $n > N$ for some N. We can then use the first half of the proof for the smallest n ■

7.3 $(S, \succeq) = (X \times Y/\sim)$ as a commutative mean groupoid

We have given $(X \times Y, \succsim)$ and $S = X \times Y/\sim$ and we can define $\circ$ on S by

Definition 93 $(a \circ b)$

$$\begin{gathered} a \circ b = c \\ \textit{if there exists} \\ (x_a, y_a) \in a \in S \textit{ and } (x_b, y_b) \in b \in S \\ \textit{such that} \\ (x_a, y_b) \sim (x_b, y_a) \in c \in S \end{gathered}$$

A.2 and A.3 of the axioms for a commutative mean groupoid will hold for $(S, \succeq, \circ)$ *but the other axioms will not hold in general. We define* $\circ$ *on* X *by*

$$x_1 \circ x_2 = x_3$$
$$\textit{if}$$
$$(x_1, y_0) \in a, (x_2, y_0) \in b, (x_3, y_0) \in c$$

and $a \circ b = c$, *for some* y_0. $\circ$ *on* Y *is defined analogously.*

$(X, \succeq, \circ) \times (Y, \succeq, \circ)$ or $(X \times Y, \succeq, \circ)$ will denote $X \times Y$ with an order defined by $(x, y) \succeq (x', y')$ if $x \succeq x'$ and $y \succeq y'$, and with an algebraic operation defined by $(x, y) \circ (x', y') = (x \circ x', y \circ y')$. So $(X \times Y, \succeq, \circ)$ is a product mean groupoid, see definition 92 page 62.

Definition 94 $(X \times Y, \succeq, \circ)$ *is homomorphic with* $(S, \succeq, \circ)$ *if*

$$[\textit{for } (x, y) \in s, (x', y') \in s', \textit{ and } (x'', y'') \in s'']$$
$$(x, y) \succeq (x', y') \textit{ implies } s \succeq s'$$
$$\textit{and}$$
$$(x, y) \circ (x', y') = (x'', y'') \textit{ implies } s \circ s' = s''.$$

It is the purpose of this section to prove

Theorem 19 *Factor independence, c.3 and T imply that* $(S, \succeq, \circ)$, $(X, \succeq, \circ)$ *and* $(Y, \succeq, \circ)$ *are commutative mean groupoids.* $(X, \succeq, \circ) \times (Y, \succeq, \circ)$ *is homomorphic with* $(S, \succeq, \circ)$.

Proof. A.1. It is an obvious consequence of R that $a \circ b$, $(x_1 \circ x_2)$ and $(y_1 \circ y_2)$ - if defined - is independent of the choice of $(x_a, y_a) \in a$, (y_0 and x_0) and T implies R according to lemma 14. $a \circ b$ is not necessarily defined for all $a, b \in S$ but according to theorem 14 if the axioms for a commutative mean groupoid hold locally, $\circ$ can in a unique way be extended to all of S. It is therefore in the following enough to check the conditions where $a \circ b$ is defined.

A.2 and A.3. These two conditions follow trivially from the definition.

A.4. To prove A.4 we shall need a definition and a lemma ∎

Definition 95 *Let* $s_1 \prec s_2 \prec s_4$ *and* $s_1 \prec s_3 \prec s_4$ *be elements from* S *and* $(x_a, y_a) \in s_1, (x_a, y_b) \in s_2, (x_b, y_a) \in s_3, (x_b, y_b) \in s_4$, *we shall say that* S *has the diagonal property or just D if whenever* $s_1 \circ s_4$ *is defined,* $s_2 \circ s_3$ *is defined and* $s_1 \circ s_4 = s_2 \circ s_3$.

D means that if the mean of the smallest and the largest of the four corners of a rectangle is defined, then the mean of the other two corners is defined and the two means are equal.

Lemma 15 *Factor independence, c.3, and T imply D.*

Proof. We assume that $s_1 \circ s_4$ is defined so there exists $(x_2, y_1) \in s_1$ and $(x_4, y_3) \in s_4$ such that $(x_2, y_3) \sim (x_4, y_1) \in s_1 \circ s_4$. Assume $x_a \precsim x_2$, $x_b \precsim x_4$, and $s_3 \succeq s_2$ (other case can be proved by analogy). Then R implies that there exists $(x_1, y_2) \in s_1$, such that $(x_1, y_3) \in s_2$, and $(x_4, y_2) \in s_3$. Define x_3 by $(x_3, y_1) \in s_2$, now T applied to $(x_1, y_2) \sim (x_2, y_1)$ and $(x_1, y_3) \sim (x_3, y_1)$ gives $(x_2, y_3) \sim (x_3, y_2)$, but $(x_2, y_3) \sim (x_4, y_1)$ so $(x_3, y_2) \sim (x_4, y_1) \in s_2 \circ s_3 = s_1 \circ s_4$ ■

We shall now continue the proof of theorem 19 by proving

Lemma 16 *Factor independence, c.3, T, and D imply A.4.*

Proof. Let

$$\begin{array}{ll} (x_a, y_b) \sim (x_b, y_a) \in a \circ b, & \\ (x_c, y_a) \sim (x_a, y_c) \in a \circ c & \text{and} \\ (x_b, y_d) \sim (x_d, y_b) \in b \circ d & \end{array}$$

Then using T

$$\begin{array}{ll} (x_b, y_c) \sim (x_c, y_b) \in b \circ c, & \\ (x_a, y_d) \sim (x_d, y_a) \in a \circ d & \text{and} \\ (x_c, y_d) \sim (x_d, y_c) \in c \circ d & \end{array}$$

now

$$\begin{array}{ll} (x_a, y_b) \in a \circ b & (x_a, y_c) \in a \circ c, \\ (x_d, y_b) \in b \circ d & (x_d, y_c) \in c \circ d \end{array}$$

are the four corners of a rectangle and D gives

$$(a \circ b) \circ (c \circ d) = (a \circ c) \circ (b \circ d)$$

M.1. This axiom follows from factor independence.

M.2. To prove M.2 let $a \succ b \succ c$. Construct sequences of points

$$x_a^1, x_a^2, \cdots ; y_a^1, y_a^2, \cdots$$
$$\text{such that}$$
$$(x_a^1, y_a^1) \in a,$$
$$(x_a^1, y_c) \sim (x_c, y_a^1) \in a \circ c$$
$$(x_c, y_c) \in c$$
$$\text{and}$$
$$(x_a^n, y_a^n) \sim (x_a^{n-1}, y_c) \sim (x_c, y_a^{n-1}) \text{ for all } n.$$

Now $x_a^n \downarrow x_a^0; y_a^n \downarrow y_a^0$ and $(x_a^0, y_c) \sim (x_a^0, y_a) \sim (x_c, y_a^0)$ so $x_a^0 = x_c$ and $y_a^0 = y_c$. This, however, means that $a \circ nc \prec b$ for $n > N$ for some N, analogously for $na \circ c \succ b$ for $n > N'$ and M.2 holds ∎

We have now proved the first part of theorem 19 and we know from theorem 13 page 55 that there exists a strictly monotonic, continuous function $f : X \times Y \to \mathbb{R}$ such that

$$f(a \circ b) = \frac{1}{2}(f(a) + f(b))$$

Simple cancelation shows that this result combined with D proves

Lemma 17 *Let*

$$(x_a, y_a) \in a, (x_a, y_b) \in b, (x_b, y_a) \in c$$
$$\textit{and}$$
$$(x_b, y_b) \in d,$$
$$\textit{then}$$
$$f(a) + f(d) = f(b) + f(c).$$

Proof. (Of theorem 19 continued) Choose now $f(a) = f(x_a, y_a) = 0$ in lemma 17. We shall then get $((x, y) = (x_b, y_b))$ $f(d) = f(x, y) = f_1(x) + f_2(y)$ where $f_1(x) = f(x, y_a)$ and $f_2(y) = f(x_a, y)$.

The function f is continuous in the order topology on S and therefore (theorem 8) continuous in the product topology on $X \times Y$. This obviously implies that f_1 and f_2 are continuous (in the order topologies on X and Y respectively).

We can now conclude the proof of theorem 19 by noting that

$$x_1 \circ x_2 = f_1^{-1}\left(\frac{1}{2}(f_1(x_1) + f_1(x_2))\right)$$

and analogously on Y. f_1 and f_2 are thus isomorphisms.

$$(f_1 : (X, \succeq, \circ) \to (f_1(X), \geqq, \circ) \text{ and } f_2 : (Y, \succeq, \circ) \to (f_2(Y), \geqq, \circ))$$

It is then trivial to check that $\chi : X \times Y \to S$ defined by $\chi(x,y) = a$ for $(x,y) \in a$ is a homomorphism ■

7.4 $f(x,y) = f_1(x) + f_2(y)$

As a part of the proof of theorem 19 we have proved

Theorem 20 *Assume factor independence, c.3, and T for* $(X \times Y, \succsim)$, *then there exist strictly monotonic and continuous* f, f_1, f_2 *such that*

$$f(x,y) = f_1(x) + f_2(y)$$

It is a trivial consequence of the preceding proofs that theorem 20 can be generalized by

Corollary 4 *Assumption T in theorem 20 can be replaced the assumption that T holds in a neighborhood of any point (neighborhood in the product topology).*

7.5 The functional equation $F(x,y) = g^{-1}(f_1(x) + f_2(y))$

This section will contain the results from the earlier sections for the special case where $\succsim$ on $X \times Y$ is derived from a function F defined on $X \times Y$ with values in a totally ordered set. We can in this way find conditions for solutions to the functional equation (7.2) page 69. We have given $X \times Y$ and $F : X \times Y \to T$ where $(T, \succeq)$ is a totally ordered space. We can then define the total preorder on $X \times Y$ by

$$(x,y) \succsim (x',y') \text{ if } F(x,y) \succeq F(x',y')$$

A total preorder can be defined on $X, (Y)$ for given $y, (x)$ as in chapter 4.

Theorem 21 *Let $F : X \times Y \rightarrow T$ where $(T, \succeq)$ is a totally ordered space. If there exist total preorders on X and Y such that X and Y are connected and F is strictly increasing and continuous in both variables, then $F(X \times Y) \subset T$ is isomorphic to an interval of $\mathbb{R}$.*

Proof. F strictly increasing for some total preorder on X and Y is equivalent to factor independence. F continuous in x and y and X and Y connected imply assumption c.3. This gives according to theorem 10 a real representation f of $(X \times Y, \succsim)$.

$$f \circ F^{-1} : F(X \times Y) \rightarrow \mathbb{R}$$

is the order-homomorphism ■

Theorem 21 shows that strictly increasing functions of two variables without much loss of generality can be assumed to be real.

Thomsen's condition becomes

$$\begin{array}{ll} F(x_0, y_1) = F(x_1, y_0) & \text{and} \\ F(x_0, y_2) = F(x_2, y_0) & \text{imply} \\ F(x_1, y_2) = F(x_2, y_1) & \end{array} \qquad \text{(Thomsen)}$$

Theorem 22 *Let X and Y be totally preordered connected sets and let $F : X \times Y \rightarrow \mathbb{R}$ be a continuous strictly increasing function for which T holds, then there exists strictly increasing continuous functions g, f_1, f_2 such that*

$$F(x, y) = g^{-1}(f_1(x) + f_2(y))$$

Proof. The assumption for theorem 20 hold and we can define $g = f \circ F^{-1}$ ■

Corollary 5 *Assumption T in theorems 21 and 22 can be replaced by the assumption that F has property T in a neighbourhood of any point*

Proof. See corollary 4 ■

7.6 Notes

7.6.1 History and further results

1. The functional equation (7.2) has a long history and appears in many different parts of mathematics. If no continuity or

monotonicity assumptions are made on g, f_1 and f_2, any F can be written in the form (7.2). Assuming differentiability de Saint-Robert (1871) [158] gave the condition

$$\frac{\partial^2}{\partial x \partial y} \log \frac{F'_x(x,y)}{F'_y(x,y)} = 0$$

for (7.2).

2. Radó (1959 and 1960) [151, 153] and Debreu (1960) [52] use the results from chapter 5 to prove results similar to theorem 20. Radó uses the solution to (7.2) to solve the bisymmetry equation (equation (6.1) page 59) from chapter 6.

3. The results from the theory of webs were generalized by Chern (1935 and 1936) [37] and [38], and was further used and developed in a series of papers by among others Chern(1966) [39], Griffiths (1976 and 1977), Chern and Griffiths (1978abc) [89, 90, 42, 40, 41], Damiano (1983) [46] and Masqué Valdés (1999)[137]. The basic results in Blaschke and Bol (1938) [30] and Debreu (1960) [52] do not use differentiability assumptions. The later theory starts with differentiable manifolds. The observation that the Thomsen and Reidemeister conditions are independence conditions in the product of the space with itself and Theorem 20 and section 7.3 strongly suggests that this special structure is not needed for the basic linearity results in this theory.

4. The idea that some kind of algebraic theory could be used to give (7.1) and (7.2) is rather obvious, see Baer (1939) [13], Aczél, Pickert and Radó (1960) [6] and Fuhrken Richter (1991) [83]. Aczél (1965) [4] gives relations between 3-nets and quasigroups (and use again the T, R, and B conditions to give properties of the quasigroups). Theorem 7.3 introduces another algebraic structure on a 3-net. The result may be new and has the obvious advantage that $a \succ a \circ b \succ b$ (for $a \succ b$) so the operation will never lead outside the elements already known to be in the set.

5. A beautiful result by Kolmogorov (1957) [112] will be mentioned here even if is not closely related to the result in this

book. There exist for any $n \geqq 2$ continuous functions

$$\psi^{pq} : [0,1] \to \mathbb{R} \, (p = 1, 2, \cdots, n; q = 1, 2, \cdots, 2n+1)$$

such that there for any continuous $f : [0,1]^n \to \mathbb{R}$ will exist continuous χ_q with

$$f(x_1, x_2, \cdots, x_n) = \sum_{q=1}^{2n+1} \chi_q \left(\sum_{p=1}^{n} \psi^{pq}(x_p) \right)$$

This shows that a continuous function of two variables can be expressed as a sum of five continuous functions of the form (7.2). The independence conditions on the graph (or condition T or B) reduces in a sense five to one.

Further references can be found in the works referred to above and in Aczél (1966) [5] footnotes 62 and 78.

Part II

Relations on Function Spaces

8

Totally preordered function spaces

8.1 Introduction

The totally preordered set in this chapter is a set $\mathcal{G}$ of functions $g : X \to Y$, where $(X, \mathcal{A})$ is a measurable set and Y an arbitrary set. Section 8.2 gives definitions and notation. As an application of the results from chapter 4 section 8.3 gives conditions for the existence of a real homomorphism on $(\mathcal{G}, \succsim)$. $\mathcal{G}$ is in section 8.4 shown to be a commutative mean groupoid $(\mathcal{G}, \succsim, \circ)$. This implies (section 8.6) the existence of functions

$$F : \mathcal{G} \to \mathbb{R}, g \succsim g' \Leftrightarrow F(g) \geqq F(g)$$
and
$$f : \mathcal{G} \times \mathcal{A} \to \mathbb{R}, f(g, A) = F(g \boxtimes_A g_0)$$

strictly monotonic in $\mathcal{G}$ and additive in $\mathcal{A}$ (only assumed to be an algebra).

If X or Y has special properties special cases and further results may be obtained.

1. The results from section 8.6 are in section 8.8 presented for the special case where the function space is a set of indicator functions (i.e. $Y = \{0, 1\}$). In this way a probability measure

   $$\alpha : \mathcal{A} \to [0, 1]$$
   such that
   $$\alpha(A) > \alpha(B) \Leftrightarrow A \succ B$$

   is obtained as an order preserving real function on an algebra.

2. The results from section 8.6 are in section 8.7 presented for the special case where X is a finite set. In this way conditions

implying a representation of a preorder on a product space by

$$\sum u_i(y_i) > \sum u_i(y_i') \Leftrightarrow y \succ y'$$

for an additive function can be given. This again is in section 8.11 used to give solutions to some functional equations.

3. There may exist a function $u : X \times Y \to \mathbb{R}$ and a measure $\mu : \mathcal{A} \to \mathbb{R}$ such that

$$f(g, A) = \int_A u(x, g(x))\, d\mu.$$

Some results giving existence of such an integral representation for Y a separable metric space will be given in chapters 11 and 12.

4. The case where Y is totally preordered and

$$(Y/\sim, \succeq, \circ)\,(\text{or } (Y/\sim_x, \succeq_x, \circ_x), x \in X)$$

is a commutative mean groupoid - for example because

$$(Y, \succsim_x, \circ_x)$$

is itself a totally preordered function space for which the independence condition holds - is important for the reformulations in chapter 14 of Fubini's theorem. $(Y/\sim_x, \succeq_x, \circ_x)$ is then isomorphic to $(\mathbb{R}_x, \geqq, \circ)$ for some $\mathbb{R}_x \subset \mathbb{R}$. Assuming consistency (definition 115 page 100) the resulting representation will be equal to an affine function on a space of real functions and is given in section 8.9. This again means that the integral representation theorems from chapters 11 and 12 can be obtained more easily for this case. An interpretation of this result is postponed to section 9.7

5. Section 8.10 treats the case where Y_x is a mean groupoid with a zero

$$(Y_x, \succsim_x, \circ_x, \Box_x)$$

where

$$(Y_x / \sim_x, \succsim_x, \circ_x, \square_x)$$

is isomorphic to

$$(\mathbb{R}_x, \geqq, \circ, 0)$$

with $0 \in \mathbb{R}_x$ for all $x \in X$. This case will be of interest in chapter 9. The consistency assumption will be made separately for positive and negative elements (y positive for $y \succ \square_x$ and y negative for $y \prec \square_x$). The resulting representation will again be an integral of the real representation of the mean groupoid on Y with respect to measures on X. An interpretation of this result is also postponed to section 9.7

6. The case where X is totally ordered - for example time - is discussed in chapter 17.

8.2 Notation and definitions

X and Y are arbitrary sets and $\mathcal{G}$ is a set of functions defined on X with values in Y. $\mathcal{A}$ is a system of subsets of X, $\succsim$ is a total preorder on $\mathcal{G}$

$$g \in \mathcal{G} \subset Y^X, \mathcal{A} \subset 2^X, graph\,(\succsim) \subset \mathcal{G} \times \mathcal{G}$$

We shall use the notation $g\,|A$ and $\mathcal{G}\,|A$ for the restrictions of g and $\mathcal{G}$ to A, the notation

$$g = \left((g_i)_{i \in I}\right)$$

for the function defined for an arbitrary index set I on

$$\bigcup_{i \in I} A_i \,(A_i \cap A_j = \emptyset \text{ for } i \neq j)$$
$$\text{by}$$
$$g\,(x) = g_i\,(x) \text{ for } x \in A_i$$

Notation 4 (mixing) *The notation $g \boxtimes_A h$ will be used for the function defined by*

$$k\,(x) = \begin{cases} g\,(x) & \textit{for } x \in A \\ h\,(x) & \textit{for } x \in A^c \end{cases}$$

Definition 96 (mixing) *g is the mixing $(g_i)_{i\in I}$ with respect to the partition $(A_i)_{i\in I}$. k is the mixing $g\boxtimes_A h$ of g and h with respect to A.*

Definition 97 (mixture) *$\mathcal{G}$ is a mixture with respect to $\mathcal{A}$ if $g_i \in \mathcal{G}$ $(i \in I)$ implies that the mixing g of $(g_i)_{i\in I}$ with respect to any finite partition $(A_i)_{i\in I}$ $(A_i \in \mathcal{A})$ is an element in $\mathcal{G}$.*

Remark 27 *It is obviously enough for $\mathcal{G}$ to be a mixture with respect to $\mathcal{A}$ that all mixings of the form $g\boxtimes_A h$ are in $\mathcal{G}$ for all $A \in \mathcal{A}$ and all $g, h \in \mathcal{G}$.*

Definition 98 ($\sigma-$mixture) *$\mathcal{G}$ is a $\sigma-$mixture with respect to $\mathcal{A}$ if $g_i \in \mathcal{G}$ $(i \in I)$ implies that the mixing g of $(g_i)_{i\in I}$ with respect to any countable partition $(A_i)_{i\in I}$ $(A_i \in \mathcal{A})$ is an element in $\mathcal{G}$.*

Definition 99 (mixture of $\mathcal{G}$) *For any $\mathcal{G} \subset Y^X$ and any $\mathcal{A} \subset 2^X$ we define $\mathcal{M}_0(\mathcal{G})$ as the smallest mixture containing $\mathcal{G}$.*

Definition 100 ($\sigma-$mixture of $\mathcal{G}$) *For any $\mathcal{G} \subset Y^X$ and any $\mathcal{A} \subset 2^X$ we define $\mathcal{M}(\mathcal{G})$ as the smallest $\sigma-$mixture containing $\mathcal{G}$.*

Remark 28 *All the results in the following assuming that a space of functions is a mixture or a $\sigma-$mixture and assuming independence can trivially be extended to arbitrary set of functions by assuming independence of a relation on $\mathcal{M}_0(\mathcal{G})$ or on $\mathcal{M}(\mathcal{G})$.*

Remark 29 *If $\mathcal{G}$ is a mixture or $\sigma-$mixture with respect to $\mathcal{A}$ we have an obvious isomorphism between*

$$\mathcal{G} \text{ and} \prod_{i\in I} \mathcal{G}\,|A_i$$

for $(A_i)_{i\in I}$ any partition of X - finite or countable - with $A_i \in \mathcal{A}$, $i \in I$.

The following lemma is trivial

Lemma 18 *If $\mathcal{G}$ is a mixture or $\sigma-$mixture with respect to $\mathcal{A}$, where $A \in \mathcal{A}$ implies $A^c \in \mathcal{A}$, $\mathcal{G}$ is also a mixture or $\sigma-$mixture with respect to the smallest algebra containing $\mathcal{A}$.*

By assuming that $\mathcal{G}$ is a mixture and regarding $\mathcal{G}$ as a product set $\left(\mathcal{G} = \prod_{i\in I} \mathcal{G}\,|A_i\right)$, we can repeat the definitions of independence from chapter 4.

Definition 101 (independence with respect to A) *$(\mathcal{G}, \succsim)$ is independent with respect to $A \in \mathcal{A}$ if $(\mathcal{G}\,|A\,, \succsim_{A,g|A^c})$ is independent of $g\,|A^c$.*

Definition 102 (independence with respect to $\mathcal{A}$) *$(\mathcal{G}, \succsim)$ is independent with respect to $\mathcal{A}$ if it is independent with respect to A for all $A \in \mathcal{A}$.*

Definition 103 (connected) *$(\mathcal{G}, \succsim)$ is connected with respect to $\mathcal{A}$ if $\mathcal{G}\,|A \times \{h\,|A^c\}$ is connected in the relative order topology. (See assumption c.3 page 33) for all $A \in \mathcal{A}$ and all $h \in \mathcal{G}$.*

Definition 104 (null set) *A subset $A \subset X$ is a null set (with respect to $\succsim$) if for all $g, h \in \mathcal{G}, g \sim h \boxtimes_A g$.*

Remark 30 *$\emptyset$ is obviously a null set.*

Definition 105 (atom) *A subset $A \subset X$ is an atom (with respect to $\succsim$) if A is not a null set and $B \subset A \in \mathcal{A}, B \in \mathcal{A}$ implies that B or $A \setminus B$ is a null set.*

Definition 106 (continuous at A) *$\succsim$ is continuous at a set A if $A_i \to A$ implies $h \boxtimes_{A_i} g \to h \boxtimes_A g$ for all $g, h \in \mathcal{G}$.*

Definition 107 (continuous) *Continuous at $\emptyset$ is called continuous.*

Definition 108 (continuous with respect to measure) *$\succsim$ is continuous with respect to a measure μ if the null sets of μ are null sets for $\succsim$.*

8.3 Real order homomorphisms

Theorem 23 *Let $(\mathcal{G}, \succsim)$ be an independent and connected mixture with respect to a partition $\mathcal{A} = \left((A_i)_{i \in I}\right)$ with more than one A_i non-null. If I is finite or $\succsim$ continuous and I countable, then there exists real order homomorphisms f and $(f_i)_{i \in I}$ on $\mathcal{G}$ and $(\mathcal{G}\,|A_i)_{i \in I}$ such that $f(g) = f_0\left(\left(f_i\left(g\,|A_i\right)\right)_{i \in I}\right)$ for some continuous strictly monotonic f_0.*

Proof. $\mathcal{G} = \prod_{i \in I} \mathcal{G}\,|A_i$ so theorem 10 and its corollary applies ■

Many function spaces are of course separable and connected in some topology, and if we assume that this topology is finer than the order topology on $(\mathcal{G}, \succsim)$, we can use corollary 2 page 23 to get a real order homomorphism. No examples of this kind of theorems shall, however, be given here.

A special case will be that there is a preorder relation on Y possibly different for different $x \in X$, so if $(Y, \succsim_x)_{x \in X} \subset Y^X$ can be represented by utility functions $u : X \times Y \to \mathbb{R}$ the space $(\mathcal{G}, \succsim)$ can be replaced by

$$\mathcal{U}(\mathcal{G}) = \widetilde{\mathcal{G}} = \left\{ \widetilde{g} \in \mathbb{R}^X \,|\widetilde{g}(x) = u(x, g(x)), g \in \mathcal{G} \right\} \subset \mathbb{R}^X$$
$$\left(\widetilde{\mathcal{G}}, \geq\right) \text{ defined by } \widetilde{g} \geq \widetilde{g}' \text{ if } \widetilde{g}(x) \geq \widetilde{g}'(x), \forall x \in X$$
$$\left(\widetilde{\mathcal{G}}, \succsim\right) \text{ defined by } \widetilde{g} \succsim \widetilde{g}' \text{ iff } g \succsim g'$$

If

$$g(x) \succsim_x g'(x), \forall x \in X \Rightarrow g \succsim g'$$

$\mathcal{G} \cap (Y, \succsim_x)_{x \in X}$ will be homomorphic with $\left(\widetilde{\mathcal{G}}, \geq\right)$, and $\left(\widetilde{\mathcal{G}}, \geq\right)$ will be homomorphic with $\left(\widetilde{\mathcal{G}}, \succsim\right)$ so if this space of real functions is separable in some topology finer than the order topology we can use corollary 2 to get a utility function on $\widetilde{\mathcal{G}}$ and then on $\mathcal{G}$. The utility of the function g then only depends on the values of the utility function for each x. We then have "utility sophistication" (For the "dual" concept of probability sophistication see Machina Schmeidler (1992) [132]). For X finite this is theorem 10. See also section 8.9 page 98.

The mapping $\mathcal{U}$ used above is part of the exponential functor defined for and on any sets Y, Z and any $u : Y \to Z$ and a fixed X and on Y^Z by

$$\mathcal{U}(Y) = Y^X$$
$$\mathcal{U}(g) = \widetilde{g} \text{ as defined above}$$
$$\left(g \in Y^X, \widetilde{g} \in Z^X\right)$$

8.4 The function space as a mean groupoid

The purpose of this section is to check that the two definitions of $\circ$ on $\mathcal{G}$ given below under some conditions coincide and that this implies that $(\mathcal{G}/\sim, \succeq, \circ)$ is a commutative mean groupoid and that condition T page 45 holds for $\mathcal{G}\,|A_1/\sim \times \mathcal{G}\,|A_2/\sim$.

Lemma 19 *Let* $(\mathcal{G}, \succsim)$ *be an independent and connected mixture with respect to an algebra* $\mathcal{A}$, *and let* $A_1, A_2, A_3\,(A = A_1 \cup A_2)$ *be three disjoint non-null sets from* $\mathcal{A}$. *Then Thomsen's condition holds for*

$$\mathcal{G}\,|A_1/\sim \times \mathcal{G}\,|A_2/\sim$$
and for
$$\mathcal{G}\,|A/\sim \times \mathcal{G}|\,A_3/\sim$$
and
$$\mathcal{G}\,|A_1/\sim, \mathcal{G}\,|A_2/\sim, \mathcal{G}\,|A/\sim \text{ and } \mathcal{G}\,|A \cup A_3/\sim$$

are commutative mean groupoids.

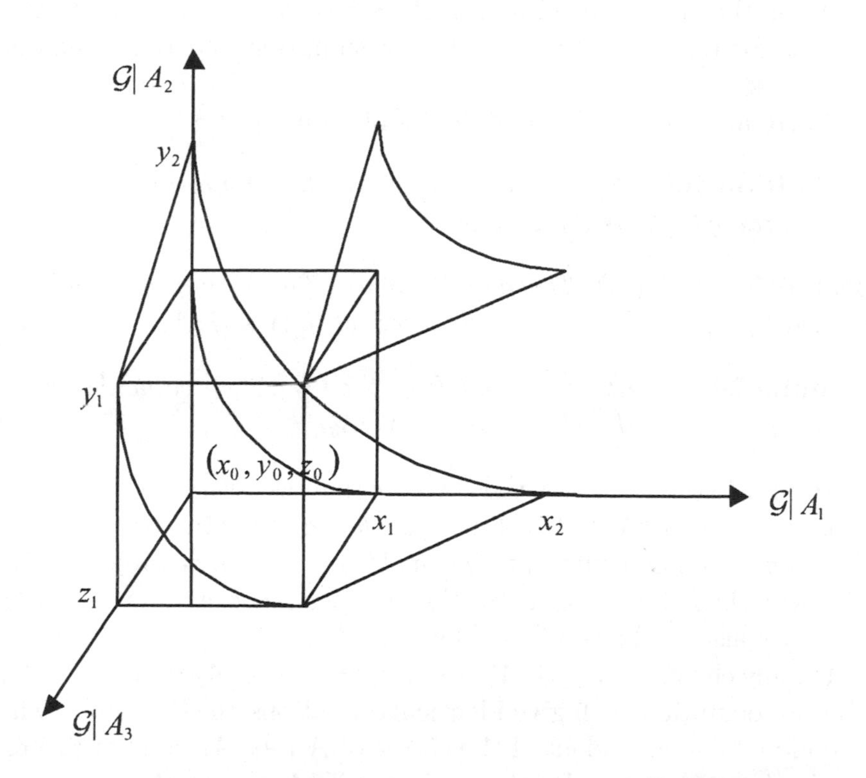

FIGURE 8.1. Thomsen holds

Proof. With the notation from figure 8.1, we chose $z_0, z_1 \in \mathcal{G}\,|A_3$ such that $(x_0, y_1, z_1) \sim (x_0, y_2, z_0)$. (This is always possible locally and theorem 14 can, if necessary, be used to extend). We then by the independence assumption have

$$(x_0, y_1, z_1) \sim (x_0, y_2, z_0) \sim (x_2, y_0, z_0) \sim (x_1, y_0, z_1)$$

and again by independence

$$(x_1, y_2, z_0) \sim (x_1, y_1, z_1) \sim (x_2, y_1, z_0)$$
so
$$(x_1, y_2) \sim (x_2, y_1)$$

and Thomsen's condition will hold in any product $\mathcal{G}\,|A_1/\sim \times \mathcal{G}\,|A_2/\sim$ where A^c is non-null. When T holds for $\mathcal{G}\,|A/\sim \times \mathcal{G}\,|B/\sim$ it will trivially hold for $\mathcal{G}\,|A'/\sim \times \mathcal{G}\,|B/\sim$ where $A \subset A'$, $A' \cap B = \emptyset$. T will therefore also hold for $\mathcal{G}\,|A/\sim \times \mathcal{G}\,|A_3/\sim$. The other conditions for theorem 19 will hold by assumption and the lemma is proved ∎

There are two possible definitions of $\circ$ on $\mathcal{G}\,|A/\sim$.

Definition 109 (A) *The definition we (for A not an atom) get by regarding $\mathcal{G}\,|A$ as the product $\mathcal{G}\,|A_1 \times \mathcal{G}\,|A_2$.*

Definition 110 (B) *The definition we (for A^c non-null) get by regarding $\mathcal{G}\,|A$ as a factor in the product $\mathcal{G}\,|A \times \mathcal{G}\,|A_3$.*

Lemma 20 *Definitions A and B of $\circ$ on $\mathcal{G}\,|A/\sim$ coincide and $\circ$ is independent of the choice of A_1, A_2, A_3.*

Proof. Figure 8.2 gives the construction necessary for the proof. The cube is constructed such that $a \circ b = c$ according to definition A. q is then chosen such that $q \in b$. This is always possible locally. We will then have $r \in c$ by the independence assumption and $a \circ b = c$ also in the product $\mathcal{G}\,|A \times \mathcal{G}\,|A_3$.

For any choice of A_1, A_2, A_3 ($A_1 \cap A_2 = \emptyset, A_1 \cup A_2 = A, A_3 \subset A^c$) the two definitions will give identical operations on $\mathcal{G}\,|A$, and $\circ$ will therefore be independent of the choice of A_1, A_2, A_3. Similar proofs easily shows that the definition of $\circ$ on $\mathcal{G}\,|A$, where A is an atom also is independent of the choice of A_3, and that the definition of $\circ$ on $\mathcal{G}$ is independent of the choice of A_1 ∎

These results can be summarized by

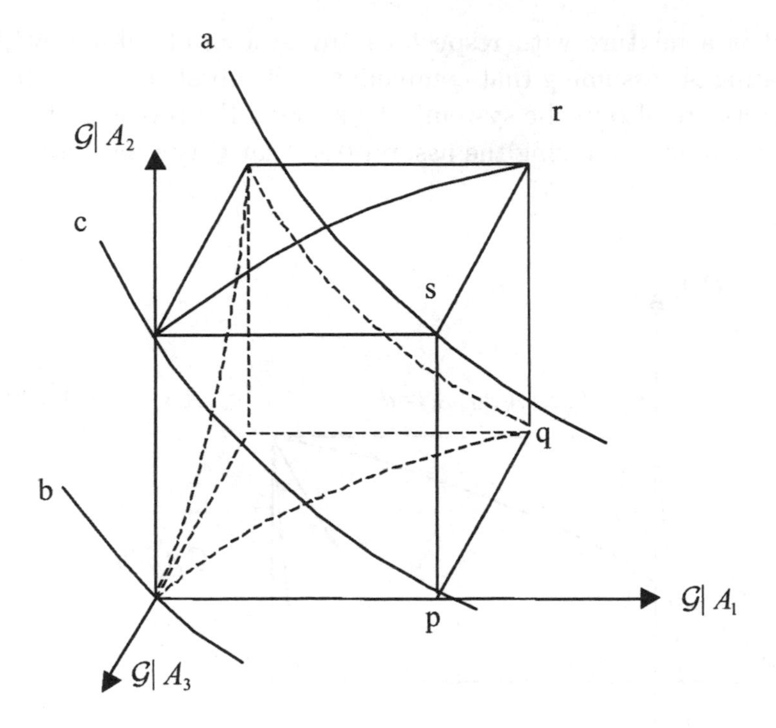

FIGURE 8.2. Definition A = Definition B

Theorem 24 *Let* $(\mathcal{G}, \succsim)$ *be an independent and connected mixture with respect to an algebra* $\mathcal{A}$ *with more than two disjoint non-null sets. Then* $(\mathcal{G}/\sim, \succeq, \circ)$ *and* $(\mathcal{G}\,|A/\sim, \succeq, \circ)$ *for all* $A \in \mathcal{A}$ *are commutative mean groupoids.* $(\mathcal{G}\,|A_1/\sim, \succeq, \circ) \times (\mathcal{G}\,|A_2/\sim, \succeq, \circ)$ *is homomorphic with* $(\mathcal{G}\,|A/\sim, \succeq, \circ)$ $(A_1 \cap A_2 = \emptyset, A_1 \cup A_2 = A)$.

Notation 5 ($\mathcal{G}$ as a mean groupoid) *It will be convenient to use the notation* $(\mathcal{G}, \succsim, \circ)$ *and to call the space of functions* $\mathcal{G}$ *with the total preorder* $\succsim$ *and with the midpoint operation* $\circ$ *defined on* $\mathcal{G}\,|A/\sim$ *for all* $A \in \mathcal{A}$ *covered by this theorem a (function space commutative) mean groupoid.*

8.5 Minimal independence assumptions

Lemma 18 shows that the assumption that $\mathcal{G}$ is a mixture with respect to an algebra can be replaced by an assumption saying that it is a mixture with respect to any system of subsets of X generating $\mathcal{A}$ (assuming that complements of subsets in the system of subsets are also in the system). This solves the problem of how far one can go in relaxing the assumption that $\mathcal{G}$ is a mixture.

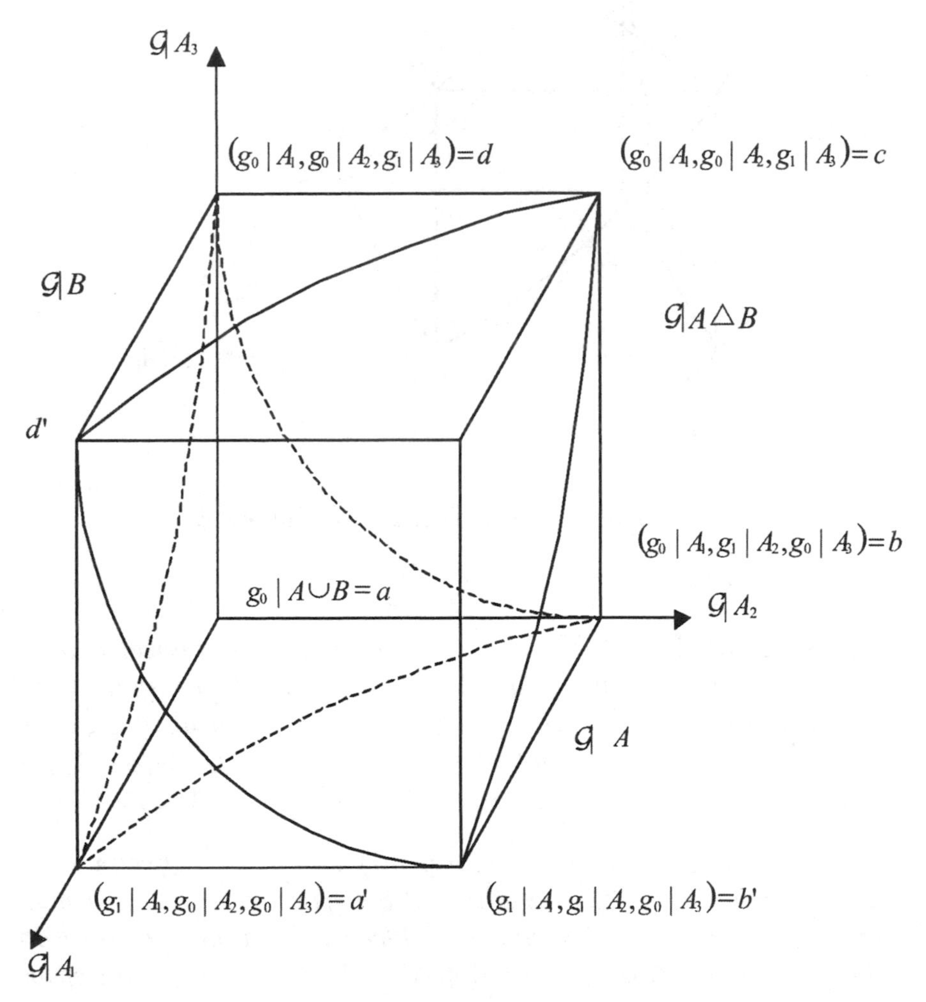

FIGURE 8.3. Minimal independence 1

The analogous problem for the independence assumption is not that trivial and is very interesting also from the point of view of most applications. Two definitions will be used.

Definition 111 (intersecting) *Two sets A_i, A_j are intersecting if $A_i \cap A_j$, $A_i \setminus A_j$ and $A_j \setminus A_i$ are non null. A family of sets $(A_i)_{i \in I}$ is intersecting if for all pairs of sets (A_i, A_j) there is a finite sequence of sets $(A_i, A_1, A_2, \cdots, A_n, A_j)$ such that consecutive pairs are intersecting.*

A basic result is

Theorem 25 *Let $(\mathcal{G}, \succsim)$ be a connected mixture with respect to an algebra $\mathcal{A}$. If $(\mathcal{G}, \succsim)$ is independent with respect to intersecting sets $A, B \in \mathcal{A}$ then $\succsim$ is independent with respect to $A \cap B, A \cup B, A \setminus B, B \setminus A$, and $A \triangle B$*[1].

Proof. Independence with respect to $A \cap B$ and $A \cup B$ is trivial. Let now $A_1 = A \cap B, A_2 = A \setminus B$ and $A_3 = B \setminus A$. Independence with respect to $A\,(B)$ means that the preorder on $\mathcal{G}\,|A_1 \times \mathcal{G}\,|A_2$ $(\mathcal{G}\,|A_1 \times \mathcal{G}\,|A_3)$ is independent of $g\,|A_3$, $(g\,|A_2)$. Let

$$
\begin{array}{lllll}
a & = & (g_0\,|A_1, g_0|\,A_2, g_0\,|A_3) & a' = & (g_1\,|A_1, g_0|\,A_2, g_0\,|A_3) \\
b & = & (g_0\,|A_1, g_1|\,A_2, g_0\,|A_3) & b' = & (g_1\,|A_1, g_1|\,A_2, g_0\,|A_3) \\
c & = & (g_0\,|A_1, g_1|\,A_2, g_1\,|A_3) & & \\
d & = & (g_0\,|A_1, g_0|\,A_2, g_1\,|A_3) & d' = & (g_1\,|A_1, g_0|\,A_2, g_1\,|A_3)
\end{array}
$$

We have to prove that the preorder on $\mathcal{G}\,|A_2 \times \mathcal{G}\,|A_3$ is independent of $g\,|A_1$.With the notation from figure 8.3, we have to prove that $d \succsim b$ implies $d' \succsim b'$ for any choice of $g\,|A_1$. Figure 8.3 shows that this is the case for $g_1\,|A_1$. $d \succ b$ and $d' \prec b'$ for some $g_1\,|A_1$implies by connectedness the existence of a g_2 such that $(g_2\,|A_1, g_0|\,A_2, g_1\,|A_3) \sim (g_2\,|A_1, g_1|\,A_2, g_0\,|A_3)$ so it is enough to prove that $d \sim b$ implies $d' \sim b'$.

The construction on figure 8.4 shows that $g' \sim c'$ where $g_2\,|A_1$is chosen in the following way: g and c are given, e is chosen such that $f \sim b$, a' is then chosen such that $a' \sim f \sim b$, then $f' \sim e \sim b', g' \sim h, c' \sim d \sim e' \sim h \sim g'$ so $c' \sim g'$. The points on $\mathcal{G}\,|A_1$ constructed by repeating this process will as in the proof of lemma 10 be dense in $\mathcal{G}\,|A_1$. $\mathcal{G}|\,A_1$ connected implies $\mathcal{G}\,|A_1$ order complete so the preorder on $\mathcal{G}\,|A_2 \times \mathcal{G}\,|A_3 = \mathcal{G}\,|A \triangle B$is independent of $g\,|A_1$ for all g. $A \setminus B = (A \triangle B) \cap A$ and $B \setminus A = (A \triangle B) \cap B$ so $(\mathcal{G}, \succsim)$ is independent also with respect to $A \setminus B$ and $B \setminus A$ ■

[1] $A \triangle B = (A \setminus B) \cup (B \setminus A)$

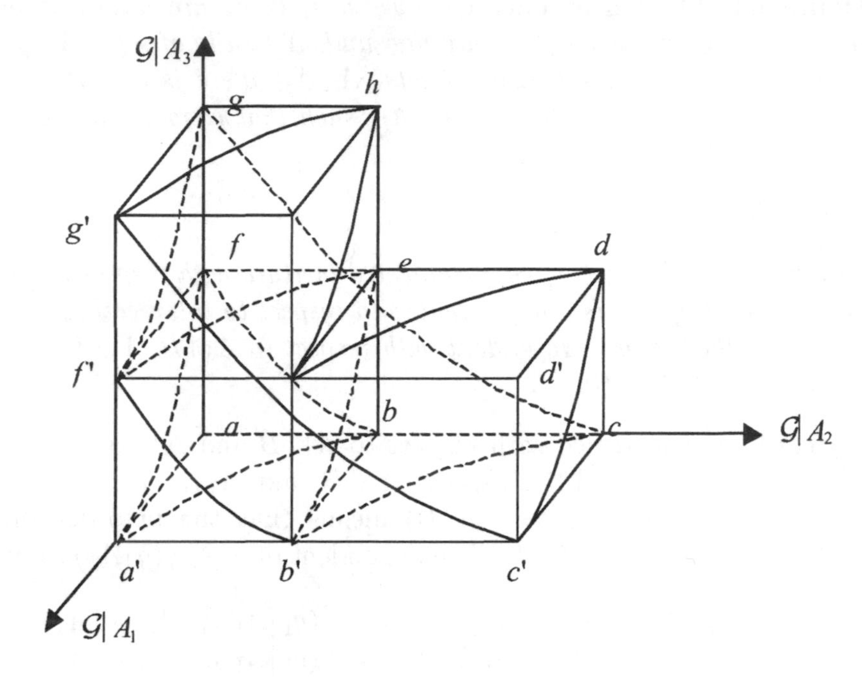

FIGURE 8.4. Minimal independence 2

Corollary 6 *Let $(\mathcal{G}, \succsim)$ be a connected mixture with respect to an algebra $\mathcal{A}$. If $(\mathcal{G}, \succsim)$ is independent with respect to $A_i, A_i^c \in \mathcal{A}$ $(i \in I,)$ where $(A_i)_{i \in I}$ is an intersecting family, then $(\mathcal{G}, \succsim)$ is independent with respect to the smallest algebra containing $(A_i)_{i \in I}$.*

Proof. Theorem 25 can be used on

$$(A_i, A_j), (A_i, A_j^c), (A_i^c, A_j) \text{ and/or } (A_i^c, A_j^c)$$

giving independence with respect to

$$A_i \cap A_j, A_i \setminus A_j, A_i \setminus A_j, \text{and } A_i^c \cap A_j^c$$

if they are non-null sets, so $(\mathcal{G}, \succsim)$ is independent with respect to all sets generated by taking complements and finite unions and intersections of sets from $(A_i)_{i \in I}$ ∎

8.6 Existence of $F : \mathcal{G} \to \mathbb{R}$ and $f : \mathcal{G} \times \mathcal{A} \to \mathbb{R}$

A combination of the results from section 8.4 and chapter 6 now easily gives the following results

Theorem 26 *Let $(\mathcal{G}, \succsim)$ be an independent and connected mixture with respect to an algebra $\mathcal{A}$ with more than two disjoint non-null sets. Then there exist functions*

$$F : \mathcal{G} \to \mathbb{R}$$
$$f : \mathcal{G} \times \mathcal{A} \to \mathbb{R}$$

such that F is a mean groupoid homomorphism. $f(\cdot, A)$ is strictly monotonic on $\mathcal{G}\,|A$ and $f(g, \cdot)$ is additive on $\mathcal{A}$.

Proof. $(\mathcal{G}, \succsim)$ is by theorem 24 a commutative mean groupoid so there exist by lemma 17 a strictly monotonic function $F : \mathcal{G} \to \mathbb{R}$ with the property $F(g) + F(h) = F(g \boxtimes_A h) + F(h \boxtimes_A g)$.

Chose now an arbitrary g_0 with $F(g_0) = 0$ and define

$$f : \mathcal{G} \times \mathcal{A} \to \mathbb{R} \text{ by } f(g, A) = F(g \boxtimes_A g_0)$$
$$\text{then}$$
$$F(g) = f(g, X) = f(g, A) + f(g, A^c)$$

$f(\cdot, A)$ is an isomorphism on $(\mathcal{G}\,|A \backslash \sim, \succeq, \circ)$ so

$$f(g, A) = f(g, A_1) + f(g, A_2), \ (A_1 \cap A_2 = \emptyset, A_1 \cup A_2 = A)$$

We now have

$$f(g, A) = \sum_{i=1}^{n} f(g, A_i)$$

where $(A_i)_{i=1}^{n}$ is any finite partition of A, so $f(g, \cdot)$ is additive. $f(\cdot, A)$ is obviously strictly monotonic on $\mathcal{G}\,|A$. (To be precise on $\mathcal{G}\,|A \times \{h\,|A^c\}$ and independent of $h\,|A^c$) ■

Remark 31 *It is easily seen from the proof that a function $\widehat{f} : \mathcal{G} \times \mathcal{A} \to \mathbb{R}$ will have the properties of f, if and only if $\widehat{f} = \alpha f + \gamma$, where $\alpha \in \mathbb{R}, \alpha > 0$, and γ is a measure on $(X, \mathcal{A}, \mu)$ which is continuous with respect to μ.*

Theorem 27 *Let $(\mathcal{G}, \succsim)$ be an independent and connected mixture with respect to a $\sigma-$algebra $\mathcal{A}$ with more than two disjoint non-null sets, and assume that $(\mathcal{G}, \succsim)$ is continuous at $\emptyset$ (with respect to a measure μ on $\mathcal{A}$). Then there exists a function*

$$f : \mathcal{G} \times \mathcal{A} \to \mathbb{R}$$

such that $f(\cdot, A)$ is strictly monotonic on $\mathcal{G}\,|A$ and $f(g, \cdot)$ is countably additive on $\mathcal{A}$ (and continuous with respect to μ).

Proof. The conditions for theorem 26 hold and an additive set function defined on a $\sigma-$algebra continuous at $\emptyset$ (with respect to a measure μ) is countably additive (and $\mu-$continuous) ■

The results so far can be illustrated in the following way:

$$\begin{array}{ccc} Y^X & & \\ \cup & & \\ (\mathcal{G}, \succsim) & \longrightarrow & (\mathcal{G}/\sim, \succeq, \circ) \\ & F & \updownarrow \\ & \searrow & \\ & & (\mathbb{R}_1, \geqq, \circ) \\ & & \cap \\ & & (\mathbb{R}, \geqq, \circ) \end{array}$$

and

$$\begin{array}{ccc} Y^A & & \\ \cup & & \\ (\mathcal{G}\,|A\,, \succsim) & \longrightarrow & (\mathcal{G}\,|A\,/\sim, \succeq, \circ) \\ & f(\cdot, A) & \updownarrow \\ & \searrow & \\ & & (\mathbb{R}_1, \geqq, \circ) \\ & & \cap \\ & & (\mathbb{R}, \geqq, \circ) \end{array}$$

and

$$(\mathcal{G}, \succsim) \times (\mathcal{A}, \uplus) \longrightarrow (\mathcal{G}/\sim, \succeq, \circ) \times (\mathcal{A}, \uplus) \longrightarrow (\mathbb{R}, \geqq, \circ, +)$$

where $\uplus$ is disjoint union and where

$$Y^X \supset (\mathcal{G}, \succsim) \longrightarrow (\mathcal{G}/\sim, \succeq, \circ)$$

says that $\mathcal{G}$ is a totally preordered space of functions and can be mapped by a surjective projection onto the set of equivalence classes, and

$$(\mathcal{G}/\sim, \succeq, \circ) \longleftrightarrow (\mathbb{R}_1, \geqq, \circ) \subset (\mathbb{R}, \geqq, \circ)$$

says that there is an isomorphism between $(\mathcal{G}/\sim, \succeq, \circ)$ and a subspace of the real numbers with the usual order and the usual midpoint.

$$(\mathcal{G}, \succsim) \times (\mathcal{A}, \uplus) \longrightarrow (\mathcal{G}/\sim, \succeq, \circ) \times (\mathcal{A}, \uplus)$$

says that there is - as before - a projection from $\mathcal{G}$ to equivalence classes in $\mathcal{G}$,

$$(\mathcal{G}, \succsim) \times (\mathcal{A}, \uplus) \longrightarrow (\mathbb{R}, \geqq, \circ, +)$$

with the usual order and the usual midpoint, and says in addition that there for each A is an isomorphism between the equivalence classes of restrictions of functions to A and a subset of the real numbers and for each $g \in \mathcal{G}$ a homomorphism between $\mathcal{A}$ with disjoint union and a subset of the real numbers with addition.

8.7 $X = \{1, 2, \cdots, n\}$ $(\prod_{i \in X} Y_i, \succsim)$

Corollary 7 *Let $\succsim$ be a total preorder on $Y_1 \times Y_2 \times \cdots \times Y_n$. Assume that $n > 2$, that $\# Y_i > 1$, that $\succsim$ is independent, that graph$\succsim$ is essential and open, and that Y_i is connected, then there exists an order homomorphism*

$$f : Y_1 \times Y_2 \times \cdots \times Y_n \to \mathbb{R}$$
such that
$$f(y_1, y_2, \cdots, y_n) = \sum u_i(y_i)$$

Proof. This is just theorem 26 for the special case considered ■

Corollary 8 *Let $\succsim$ be a total preorder on $Y_1 \times Y_2 \times \cdots \times Y_n$, and let for all y_i the preorder on $\prod_{i \neq j} Y_i$ be independent of y_i. If $(Y_i/\sim, \succeq)$ is connected in the relative topology and has more than one element for at least three i's, then there exists an order homomorphism*

$$f : Y_1 \times Y_2 \times \cdots \times Y_n \to \mathbb{R}$$
such that
$$f(y_1, y_2, \cdots, y_n) = \sum u_i(y_i)$$

Proof. Let $X = \{1, 2, \cdots, n\}$ and use corollary 6 ■

8.8 $Y = \{0,1\}, (\mathcal{A}, \succsim)$

This is the important special case where the total preorder is defined on $\mathcal{A}$, or equivalently on a class of indicator functions on X. The assumptions for a representation theorem are of course unchanged but it is useful to state the independence assumption.

Definition 112 (uncertainty space) $(X, \mathcal{A}, \succsim)$ *where* X *is an arbitrary set,* $\mathcal{A}$ *an algebra of subsets of* X*, and* $\succsim$ *a total preorder on* $\mathcal{A}$.

Definition 113 (subjective probability) *Let* $\succsim$ *be a total preorder on* $\mathcal{A}$. *An additive*

$$\alpha : \mathcal{A} \to [0,1]$$
$$\text{such that}$$
$$\alpha(A) > \alpha(B) \Leftrightarrow A \succ B$$

is called subjective probability (representing the order relation on $\mathcal{A}$*).*

Definition 114 (independence) $\succsim$ *on* $\mathcal{A}$ *is independent if for all* $A, A_1, A_2, C \in \mathcal{A}, C \subset A^c, A_1 \cup A_2 \subset A$

$$A_1 \succsim A_2 \Longleftrightarrow A_1 \cup C \succsim A_2 \cup C$$

Corollary 9 *Let* $(X, \mathcal{A}, \succsim)$ *be an uncertainty space, where* $\succsim$ *is a total preorder. Let* $\succsim$ *be independent and let* $\mathcal{A}|A$ $(A \in \mathcal{A})$ *be connected. Then there exist a subjective probability.*

Proof. Special case of theorem 26 ■

8.9 Y a commutative mean groupoid

Theorem 24 implies that $\mathcal{G}$ is a commutative mean groupoid and theorems 26 and 27 give the general representations. Special structure on Y will give extra results. Y may itself be a commutative mean groupoid, (i.e. isomorphic to a subset of the real numbers) it may be a product space mean groupoid, or it may a function space mean groupoid (8.9.1). This section covers the case where Y is a commutative mean groupoid. The special case where Y is itself a function space mean groupoid (subsection 8.9.1) leads in chapter 14 to a way of formulating Fubini's theorem. The isomorphism with the real numbers means that the results are relatively trivial results about totally preordered sets of real functions. The next section (8.10) covers the case where Y is a commutative mean groupoid with zero. These results are generalized in chapter 9 to obtain results where the mean groupoids are derived from relations on $\mathcal{G}$ and Y not assumed to be total preorders.

Let $(Y_x, \succsim_x, \circ_x)_{x \in X}$ be totally preordered spaces with a midpoint operation defined on $S_x = Y_x / \sim_x$ making $(S_x, \succeq_x, \circ_x)$ a commutative mean groupoid for all $x \in X$. There exists then for any choice of $g_0, g_1 \in \mathcal{G}$ with $g_0(x) \prec g_1(x)$ for all $x \in X$ a representation

$$\begin{gathered} F_x : Y_x \to \mathbb{R}, (F_x)_{x \in X} : (Y_x)_{x \in X} \to \mathbb{R}^X \\ F_x(g_0(x)) = 0, F_x(g_1(x)) = 1 \\ \text{with for all } x \in X \\ F_x(s) \geqq F_x(s') \Leftrightarrow s \succeq_x s' \\ \text{and} \\ F_x(s_0) = \tfrac{1}{2} F_x(s_1) + \tfrac{1}{2} F_x(s_2) \Leftrightarrow s_0 = s_1 \circ_x s_2 \\ (Y_x, \succsim_x, \circ_x)_{x \in X} \to (S_x, \succeq_x, \circ_x)_{x \in X} \overset{F_x}{\leftrightarrow} (\mathbb{R}_x, \geqq, \circ)_{x \in X} \end{gathered}$$

where $\mathbb{R}_x = F_x(\cdot)$. F_x determines a mapping $\mathcal{F}$ (the exponential functor) between functions on X with values in S_x (and therefore with values in $S = \cup S_x$) and functions on X with values in $\mathbb{R}$

$$\begin{array}{ccc} 2^{Y^X} \supset & & 2^{S^X} \supset \\ (Y_x, \succeq_x, \circ_x)_{x \in X} \cap \mathcal{G} & \overset{\chi}{\to} & (S_x, \succeq_x, \circ_x)_{x \in X} \cap \chi(\mathcal{G}) \\ & & \updownarrow \mathcal{F} \\ & & \left\{ (\mathbb{R}_x, \geqq, \circ)_{x \in X} \cap \widetilde{\mathcal{G}} \right\} \subset \\ & & 2^{\left(\mathbb{R}^X, \geqq, \circ\right)} \end{array}$$

Let the real functions $\widetilde{\mathcal{G}} = \mathcal{F}(\mathcal{G}) \subset \mathbb{R}^X$ be the image of this mapping for $\mathcal{G}$ so

$$\widetilde{g} \in \widetilde{\mathcal{G}} = \left\{ \widetilde{g} \in \mathbb{R}^X \mid \widetilde{g}(x) = F_x(g(x)), g \in \mathcal{G} \right\}$$

$\mathbb{R}^X$ and therefore $\widetilde{\mathcal{G}}$ has the an order relation $\geqq$ defined by

$$\begin{gathered} \widetilde{g_1} \geqq \widetilde{g_2} \\ \text{iff} \\ g_1(x) \succsim_x g_2(x) \ \text{ for all } x \in X \\ \Leftrightarrow \\ F_x(g_1(x)) \geqq F_x(g_2(x)) \ \text{ for all } x \in X \end{gathered}$$

and a midpoint operation defined by

$$\widetilde{g} = \widetilde{g}_1 \widehat{\circ} \widetilde{g}_2 = \tfrac{1}{2}\widetilde{g}_1 + \tfrac{1}{2}\widetilde{g}_2$$
$$\text{iff}$$
$$\widetilde{g}(x) = \tfrac{1}{2}\widetilde{g}_1(x) + \tfrac{1}{2}\widetilde{g}_2(x) \text{ for all } x \in X$$

S^X is homomorphic with $\mathbb{R}^X$. Denote $(S_x)_{x\in X}$ with the structure this implies by $\left((S_x)_{x\in X}, \widehat{\succeq}, \widehat{\circ}\right)$.

Let

$$(\mathcal{G}, \succsim, \circ)$$

where $\mathcal{G}/\sim_x \subset (S_x)_{x\in X} \subset S^X$, and $(S_x, \succeq_x, \circ_x)$ be commutative mean groupoids and

$$F : \mathcal{G} \to \mathbb{R}, f : \mathcal{G} \times \mathcal{A} \to \mathbb{R}$$

be representations for $(\mathcal{G}, \succsim, \circ)$.

Define

$$\widetilde{F} : \widetilde{\mathcal{G}} \to \mathbb{R}, \widetilde{f} : \widetilde{\mathcal{G}} \times \mathcal{A} \to \mathbb{R}$$
$$\text{by}$$
$$F(g) = \widetilde{F}(\widetilde{g}), f(g, A) = \widetilde{f}(\widetilde{g}, A)$$

The assumption which imply new properties of the representations is that this functor yields a homomorphism.

Definition 115 (consistency) $\left((Y_x)_{x\in X}, \widehat{\succsim}, \widehat{\circ}\right), \left((S_x)_{x\in X}, \widehat{\succeq}, \widehat{\circ}\right)$ *and* $(\mathcal{G}, \succsim, \circ)$ *are consistent if the maps*

$$\mathcal{G} \cap \left((S_x)_{x\in X}, \widehat{\succeq}, \widehat{\circ}\right) \to (\mathcal{G}/\sim, \succeq, \circ) \leftrightarrow (\mathbb{R}_1, \geqq, \circ) \subset (\mathbb{R}, \geqq, \circ)$$

give a homomorphism between $\mathcal{G} \cap \left((S_x)_{x\in X}, \widehat{\succsim}, \widehat{\circ}\right)$ *and* $(\mathbb{R}, \geqq, \circ)$, *i.e.*

$$\begin{array}{rlr} g_1 \widehat{\succeq} g_2 \Rightarrow g_1 \succsim g_2 & & \textit{or} \quad \widehat{\succeq} \Rightarrow \succsim \\ \widetilde{g}_0 = \frac{1}{2}\widetilde{g}_1 + \frac{1}{2}\widetilde{g}_2 \Rightarrow G_0 = G_1 \circ G_2 & & \textit{or} \quad \widehat{\circ} \Rightarrow \circ \\ \textit{for } G_0, G_1, G_2 \in \mathcal{G}/\sim, \textit{ and } g_i \in G_i, i = 0, 1, 2 & & \end{array}$$

Theorem 28 *Assume consistency, then* $\widetilde{F}$ *and* $\widetilde{f}(\cdot, A)$ *are affine for all* $A \in \mathcal{A}$ *and* $\widetilde{f}(\widetilde{g}, \cdot)$ *is additive.*

Proof. F_x a representation for the mean groupoid on S_x for all $x \in X$ means that

$$g_0(x) = g_1(x) \circ_x g_2(x)$$
$$\Rightarrow$$
$$F_x(g_0(x)) = \tfrac{1}{2} F_x(g_1(x)) + \tfrac{1}{2} F_x(g_2(x))$$
$$\Longleftrightarrow$$
$$\widetilde{g}_0(x) = \tfrac{1}{2}\widetilde{g}_1(x) + \tfrac{1}{2}\widetilde{g}_2(x)$$

Consistency means that for $g_i \in G_i \in \mathcal{G}/\sim$

$$G_0(x) = G_1(x) \circ_x G_2(x)$$
$$\Rightarrow$$
$$G_0 = G_1 \circ G_2$$

$f(\cdot, A)$ a representation for the commutative mean groupoid on $\mathcal{G}\,|A$ means that

$$f(G_1 \circ G_2, A) = \tfrac{1}{2} f(G_1, A) + \tfrac{1}{2} f(G_2, A)$$
and then
$$\widetilde{f}(\widetilde{g}_0, A) = \widetilde{f}\left(\tfrac{1}{2}\widetilde{g}_1 + \tfrac{1}{2}\widetilde{g}_2, A\right) = \tfrac{1}{2}\widetilde{f}(\widetilde{g}_1, A) + \tfrac{1}{2}\widetilde{f}(\widetilde{g}_2 A)$$

extending first to dyadic numbers and then by continuity to all $\lambda \in [0,1]$ gives finally

$$\widetilde{f}(\lambda\widetilde{g}_1 + (1-\lambda)\widetilde{g}_2, A) = \lambda\widetilde{f}(\widetilde{g}_1, A) + (1-\lambda)\widetilde{f}(\widetilde{g}_2 A)$$

so $\widetilde{f}(\cdot, A)$ is affine. $\widetilde{f}(\widetilde{g}, \cdot)$ is additive because $f(g, \cdot)$ is. The properties of $\widetilde{F}$ follows from $\widetilde{F}(\widetilde{g}) = \widetilde{f}(\widetilde{g}, X)$ ■

Definition 116 (measurable) $\mathcal{G} \subset (S_x, \succsim_x, \circ_x)_{x \in X}$ *is measurable (with respect to $\mathcal{A}$ and for a given g_0, g_1) if $\widetilde{g} \in \widetilde{\mathcal{G}} \subset \mathbb{R}^X$ are Lebesgue measurable.*

Theorem 29 *Let $\mathcal{G} \subset (S_x, \succsim_x, \circ_x)$ with the representation $F_x : Y \to \mathbb{R}$ be such that $(S_x, \succsim_x, \circ_x)_{x \in X}$ is measurable and consistent with the mean groupoid $(\mathcal{G}, \succsim, \circ)$ assumed to be continuous, then there exists a normalization of F and f and measure α on $\mathcal{A}$ such that*

$$F(g \boxtimes_A g_0) = f(g, A) = \widetilde{f}(\widetilde{g}, A) = \int_A F_x(g(x))\, d\alpha \qquad (*)$$

Proof. Normalize F and f by

$$f(g_0, A) = F(g_0) = 0, \text{ and } F(g_1) = 1$$

α is defined by $\alpha(A) = F(g_1 \boxtimes_A g_0)$. Then F and f has to be as determined by $(*)$ first for simple functions with values 0 or 1, then by consistency for simple functions with binary numbers as values, and then by continuity extended to all measurable functions[2] ■

Remark 32 *$\mathcal{G}$ may consist of all functions S^X and then of course $\mathcal{G}$ could not be measurable. In contrast with the results in chapters 11 and 12 we can always get large classes of functions with an integral representation, but one of those classes may not contain all the functions. Different choices of g_0, g_1 will result in different classes of measurable functions, and for any choice of g_0, g_1 will we get a set of measurable functions for which the $(*)$ representation will hold.*

8.9.1 $(\mathcal{H}, \succsim, \circ)$

The case where S is itself a set of equivalence classes from a totally preordered independent space of functions - and therefore a function space mean groupoid - is important as a way of formulating Fubini's theorem (chapter 14) and therefore as a starting point for generalizing Fubini's theorem to the case where the basic relations are no longer total and transitive. It means that the original space is a space of functions of two variables $\left(\subset Z^{XY}\right)$. The notation will be $\mathcal{H}_x \subset Z^Y$ $(Y, \mathcal{B})$ $\mathcal{H} = \cup_x \mathcal{H}_x, \mathcal{G} \subset (\mathcal{H}_x)_{x \in X} \subset X \times \mathcal{H}$ and with the representations

$$\Phi : \mathcal{G} \to \mathbb{R}$$
$$\varphi : \mathcal{G} \times \mathcal{B} \to \mathbb{R}$$

in particular

$$\Phi(x, h) \geqq \Phi(x, h') \Longleftrightarrow h(x, \cdot) \succsim_x h'(x, \cdot)$$

Except for the form of the representation this is exactly the same problem as in the earlier part of this section. The consistency assumption is the same and the measurability assumption also the

[2] For the measure theory used see for example §§ 23-25, pp 95-102, in Halmos (1961) [93].

same. We can therefore just repeat the theorems with the changed form of the representation on the mean groupoid for any $x \in X$.

Theorem 16 is unchanged. Theorem 17 changes slightly to

Theorem 30 *Let*

$$(\mathcal{H}_x, \precsim_x, \circ_x)_{x \in X}$$

with the representation

$$\Phi : X \times \mathcal{H} \to \mathbb{R}$$

be measurable and consistent with the mean groupoid $(\mathcal{G}, \precsim, \circ)$ *assumed to be continuous, then there exists a normalization of* F *and* f *and measure* α *on* $\mathcal{A}$ *such that*

$$F(g \boxtimes_A g_0) = f(g, A) = \widetilde{f}(\widetilde{g}, A) = \int_A \Phi(x, h)\, d\alpha = \int_A \varphi(x, h, Y)\, d\alpha, \text{ with } \varphi(x, h, \cdot) \text{ additive in } \mathcal{B} \qquad (**)$$

Remark 33 *Despite all the notation the results in this last section are of course trivial.*

8.10 Y a commutative mean groupoid with zero

Let now $(Y_x, \precsim_x, \circ_x, \square_x)$ be a commutative mean groupoid with a zero for all $x \in X$, and $(\mathcal{G}, \precsim, \circ, \square)$ a function space mean groupoid also with a zero. Let the representations be as in section 8.9 with the normalization

$$g_0(x_0) = \square_x \text{ and } F(g_0) = 0$$

(so $\Phi(x, g_0(x)) = 0$).

Define

$$Y_x^+ = \{y \in Y \,|\, y \succ_x \square_x\}$$
$$Y_x^- = \{y \in Y \,|\, y \prec_x \square_x\}$$

$$\mathcal{G}^+ = \{g \in \mathcal{G} \,|\, g(x) \succ_x \square_x, x \in X\}$$
$$\mathcal{G}^- = \{g \in \mathcal{G} \,|\, g(x) \prec_x \square_x, x \in X\}$$
$$\mathcal{G}^0 = \{g \in \mathcal{G} \,|\, g(x) \sim_x \square_x, x \in X\}$$

Any g for which

$$\begin{array}{l} X^+ = \{x \in X \,|g(x) \succ_x \Box_x\} \in \mathcal{A} \\ X^- = \{x \in X \,|g(x) \prec_x \Box_x\} \in \mathcal{A} \\ X^0 = \{x \in X \,|g(x) \sim_x \Box_x\} \in \mathcal{A} \end{array}$$

can then be uniquely written

$$g = \left(g^+ \left|X^+, g^-\right| X^-, g^0 \left|X^0\right.\right)$$

where $g^+ \in \mathcal{G}^+$, $g^- \in \mathcal{G}^-$and $g^0 \in \mathcal{G}^0$. Obviously $F(\mathcal{G}^+) > 0, F(\mathcal{G}^-) < 0$ and $F(\mathcal{G}^0) = 0$

Definition 117 (consistency) $(Y_x, \succsim_x, \circ_x, \Box_x)_{x\in X}$ *and* $(\mathcal{G}, \succsim, \circ, \Box)$ *are consistent if*

$$g_0 \in \Box, \tag{1}$$

$$\begin{array}{l} (Y_x^+, \succsim_x, \circ_x) \text{ and } (\mathcal{G}^+, \succsim, \circ) \\ [(Y_x^-, \succsim_x, \circ_x) \text{ and } (\mathcal{G}^-, \succsim, \circ) \end{array} \tag{2}$$

are consistent according to definition 115 page 100, and for all $g \in \mathcal{G}$

$$\begin{array}{l} X^+ = \{x \in X \,|g(x) \succ_x \Box_x\} \in \mathcal{A} \\ X^- = \{x \in X \,|g(x) \prec_x \Box_x\} \in \mathcal{A} \\ X^0 = \{x \in X \,|g(x) \sim_x \Box_x\} \in \mathcal{A} \end{array} \tag{3}$$

Theorem 31 *Assume consistency, then* $\widetilde{F}$ *and* $\widetilde{f}(\cdot, A)$ *are affine on* $\mathcal{G}^+$ *and on* $\mathcal{G}^-$ *for all* $A \in \mathcal{A}$ *and* $\widetilde{f}(\widetilde{g}, \cdot)$ *is additive.*

Proof. The proof of theorem 16 can be applied with the obvious changes ■

Remark 34 *The functions* $\widetilde{F}$ *and* $\widetilde{f}(\cdot, A)$ *defined on* $\mathcal{G}^+ \cup \mathcal{G}^-$, *can be uniquely extended to all functions in* $\mathcal{G}$ *for which the* X^+ *and* X^- *defined above are elements in* $\mathcal{A}$ *by*

$$\begin{array}{c} \widetilde{f}(g, A) = \widetilde{f}(g^+, X^+ \cap A) + \widetilde{f}(g^-, X^- \cap A) \\ \text{and } \widetilde{F} = \widetilde{f}(\cdot, X) \end{array}$$

Theorem 32 *Let $(Y_x, \succsim_x, \circ_x, \square_x)$ be a mean groupoid with a zero with the representation*

$$F_x : Y_x \to \mathbb{R}$$

Where

$$\mathcal{G} \subset (Y_x)_{x \in X} \subset Y^X$$

is measurable and consistent with the mean groupoid

$$(\mathcal{G}, \succsim, \circ, \square)$$

assumed to be continuous, then there exists a normalization of F and f and measures α, β on $\mathcal{A}$ such that

$$\begin{gathered} F(g \boxtimes_A g_0) = f(g, A) = \widetilde{f}(\widetilde{g}, A) = \\ \int_{X^+ \cap A} F_x(g(x))\, d\alpha + \int_{X^- \cap A} F_x(g(x))\, d\beta = \\ \int_A F_x{}^+(g(x))\, d\alpha + \int_A F_x{}^-(g(x))\, d\beta = \\ \int_A U(x, g(x))\, d\gamma \end{gathered} \qquad (***)$$

where

$$U : X \times Y \to \mathbb{R}$$

is defined by

$$U(x, g(x)) = \begin{cases} F_x(g(x)) \frac{d\alpha}{d\gamma} & \text{on } X^+ \\ F_x(g(x)) \frac{d\beta}{d\gamma} & \text{on } X^- \end{cases}$$

and γ is any measure such that α and β arc absolutely continuous with respect to γ.

Proof. Again the same proof as for theorem 17 with the obvious modifications can be used. The measurability assumption implies that X^+ and X^- are measurable. ■

Remark 35 *$F, f, \widetilde{f}$ are determined up to one positive constant and F_x, α, β are - given F - determined up to a Radon Nikodým derivative*

Remark 36 *$U(x, \cdot)$ is a mean groupoid representation on*

$$(Y_x^+, \succsim_x, \circ_x, \square_x)$$

and on

$$(Y_x^-, \succsim_x, \circ_x, \square_x)$$

but not on all of Y. There is a kink in zero.

8.10.1 $(\mathcal{H}, \succsim, \circ_x, \Box_x)$

Also in this case it will be useful to make explicit the case where S is itself a set of equivalence classes from a totally preordered independent space of functions - and therefore a function space mean groupoid, now with a zero. It means that the original space is a space of functions of two variables $\left(\subset Z^{XY}\right)$. The notation will - as before - be $\mathcal{H}_x \subset Z^Y$ $(Y, \mathcal{B})$ $\mathcal{H} = \cup_x \mathcal{H}_x, \mathcal{G} \subset (\mathcal{H}_x)_{x \in X} \subset X \times \mathcal{H}$ and with the representations

$$\begin{gathered} \Phi : \mathcal{G} \to \mathbb{R} \\ \varphi : \mathcal{G} \times \mathcal{B} \to \mathbb{R} \end{gathered}$$

in particular

$$\begin{gathered} \Phi(x, h) \geqq \Phi(x, h') \Longleftrightarrow h(x, \cdot) \succsim_x h'(x, \cdot) \\ \Phi(x, \Box_x) = 0 \end{gathered}$$

Except for the form of the representation this is exactly the same problem as in the earlier part of this section. The consistency assumption is the same and the measurability assumption also the same. We can therefore just repeat the theorems with the changed form of the representation on the mean groupoid for any $x \in X$.

Theorem 31 is unchanged. Theorem 32 changes slightly to

Theorem 33 *Let*

$$(\mathcal{H}_x, \succsim_x, \circ_x, \Box_x)_{x \in X}$$

with the representation

$$\Phi : X \times \mathcal{H} \to \mathbb{R}$$

be measurable and consistent with the mean groupoid $(\mathcal{G}, \succsim, \circ, \Box)$ assumed to be continuous, then there exists a normalization of F and f and measure α on $\mathcal{A}$ such that

$$\begin{gathered} F(g \boxtimes_A g_0) = f(g, A) = \widetilde{f}(\widetilde{g}, A) = \\ \int_A \Phi^+(x, h)\, d\alpha + \int_A \Phi^-(x, h)\, d\beta = \\ \int_A \varphi^+(x, h, Y)\, d\alpha + \int_A \varphi^-(x, h, Y)\, d\beta \\ \textit{with } \varphi(x, h, \cdot) \textit{ additive in } \mathcal{B} \end{gathered} \qquad (****)$$

8.11 Related functional equations

Some of the results in this chapter can - like the results in chapter 7 - by a rather simple translation be presented as solutions to functional equations.

Theorem 34 *Let X, Y, Z be arbitrary sets, $(T, \succeq)$ a totally ordered connected set, and $F : X \times Y \times Z \to T$ a function such that*

$$\begin{array}{l} F(X \times \{y\} \times \{z\}) \\ F(\{x\} \times Y \times \{z\}) \\ F(\{x\} \times \{y\} \times Z) \end{array} \tag{8.1}$$

are connected for all (x, y, z), and such that there exist functions F_1, F_2, F_3, F_4 with a property analogous to (8.1) and totally ordered sets T_1 and T_2

$$\begin{array}{ll} F_1 : X \times Y \to T_1 & F_3 : T_1 \times Z \to T \\ F_2 : Y \times Z \to T_2 & F_4 : X \times T_2 \to T \end{array}$$

with F of the form

$$F(x, y, z) = F_3(F_1(x, y), z) = F_4(x, F_2(y, z)) \tag{8.2}$$

then F is of the form

$$F(x, y, z) = f_0(f_1(x) + f_2(y) + f_3(z))$$

where f_1, f_2, f_3 are real functions and $f_0 : \mathbb{R} \to T$ is strictly monotonic.

Proof. A total preorder is in the obvious way induced on $X \times Y \times Z$. (A) implies that the connectedness assumptions in theorem 26 will hold (with $X = \{1, 2, 3\}$). (8.2) implies that the preorder is independent with respect to $(\{1, 2\}\ \{2, 3\})$. It is then by theorem 25 also independent with respect to all subsets of $\{1, 2, 3\}$. Theorem 26 and corollary 8 now gives the isomorphism

$$f : ((X \times Y \times Z) / \sim) \to \mathbb{R}_1$$

(where $\mathbb{R}_1$ is a connected subset of $\mathbb{R}$ and

$$f(x, y, z) = f_1(x) + f_2(y) + f_3(z))$$

f_0 is then defined by $f_0 = F \circ f^{-1}$ ■

Another example is

Theorem 35 *Let $F : Y_1 \times Y_2 \times \cdots \times Y_n \to \mathbb{R}$ be a real function of n variables $(n > 2)$. If*

$$\{x \in \mathbb{R} \,|x = F(y_1, y_2, ..., y_n), y_i \in Y_i\}$$

is connected for all i and all $y_j \in Y_j$ $(j \neq i)$, and

$$\begin{array}{r} F((y_1, y_2, ..., y_n)) = \\ F_1((g_1(y_1, y_2), \cdots, y_n)) = \\ F_2((y_1, g_2(y_2, y_3), \cdots, y_n)) = \ldots = \\ F_{n-1}((y_1, y_2, ..., g_{n-1}(y_{n-1}, y_n))) \end{array}$$

for some real functions $(g_i)_{i=1,2,\cdots,n-1}$. Then

$$F((y_1, y_2, ..., y_n)) = f_0\left(\sum f_i(y_i)\right)$$

Proof. The proof of theorem 34 can be copied almost word by word ■

8.12 Notes

1. The definition of the mean groupoid operation on a totally preordered function space and the use of the Aczél Fuchs theorem (theorem 13 page 55) to obtain theorem 26 may be new. The function $f : \mathcal{G} \times \mathcal{A} \to \mathbb{R}$ will have many of the properties of integrals, and when the Hahn decomposition can be applied (theorem 27) a lattice structure can be defined on $\mathcal{G}$ such that $F(g) = f(g, X)$ becomes a valuation on $\mathcal{G}$. The results of Alfsen (1963) [7] can then be used to get results comparable to the results in chapter 11.
   The results in section 8.5 are implicit in some results on functional equations, see Aczél (1966) [5], Radó (1959) [152], Leontief (1947a,b) [123, 122] and can be found in Gorman (1968) [87]. For further references on the functional equations in section 8.11 see Aczél (1966) [5] and Radó (1959b) [152].

2. For many applications would it be of interest to have results where $\mathcal{G}$ was not a mixture (continuous functions or trajectories from a control problem). Only few (see Zhou (1999) [190]) results exist and it appears difficult to formulate conditions like the independence assumption.

3. The independence assumption is hard to accept for many applications of theorem 26 in economic theory, and it would be very important to have generalizations of the theorem. The special functions we may hope to get could be of at least two different forms

   (a) It is well-known (Aczél (1966) [5] theorem 5.3.1.2.) that

$$F(x_1+t, x_2+t, \cdots, x_n+t) = F(x)+t$$
and
$$F(x_1u, x_2u, \cdots, x_nu) = F(x)u, u \neq 0$$

   characterize functions of the form

$$F(x_1, x_2, \cdots, x_n) = \overline{x} + \sigma G\left(\frac{x_1-\overline{x}}{\sigma}, \frac{x_2-\overline{x}}{\sigma}, \cdots, \frac{x_n-\overline{x}}{\sigma}\right) \tag{8.3}$$

   where $\overline{x} = \frac{1}{n}\sum x_i, \sigma = \sqrt{\frac{1}{n}\sum(x-\overline{x})^2}$, and G an arbitrary function. For many economic applications would it be important to have (8.3) as a conclusion replaced by

$$F(x_1, x_2, \cdots, x_n) = f(\overline{y}) + G((z_1-\overline{z}), (z_2-\overline{z}), \cdots, (z_n-\overline{z}))$$
where
$$\overline{y} = \frac{1}{n}\sum y_i, y_i = g_i(x_i), \overline{z} = \frac{1}{n}\sum z_i, z_i = h_i(x_i) \tag{8.4}$$

   and $f, (g_i, h_i)$ unknown functions. (8.4) expresses that $F(x)$ (the utility in some economic interpretations) depend on the average and the distribution around the average, but the transformations $f, (g_i, h_i)$ that are needed before the averages are taken are unknown.

   (b) Another generalization of the theorem may be more useful.

$$F((y_1, y_2, ..., y_n)) = f_0\left(\sum f_i(y_i)\right)$$

   can obviously be generalized by

$$F((y_1, y_2, ..., y_n)) = f_0\left(\sum\sum f_{ij}(y_i, y_j)\right)$$

similarly

$$f : \mathcal{G} \times \mathcal{A} \rightarrow \mathbb{R}$$

can be generalized by

$$f : \mathcal{G} \times \mathcal{B} \rightarrow \mathbb{R} \tag{8.5}$$

where $\mathcal{B}$ is a system of subsets of $X \times X$ ($f(g, A_1 \times A_1)$ and $f(g, A_2 \times A_2)$ would then be the utility of g in the subsets A_1 and A_2. $f(g, A_1 \times A_2)$ would then be an interaction term. $f(\cdot, A \times A)$ should be strictly monotonic in $g\,|A$ and $f(g, \cdot)$ should be additive.
At least three different types of independence assumptions may play a role in finding conditions for (8.5).

i. The order on $\mathcal{G}\,|A$ for a given $g\,|A^c$ may be independent of $g\,|A^c$. This is the independence assumption.
ii. The order on $\mathcal{G}\,|B$ for a given $g\,|A^c (B \in \mathcal{A}, B \subset A$ is independent of $g\,|A \setminus B$, but the order on $\mathcal{G}\,|B$ may depend on $g\,|A^c$.
iii. Given disjoint $A, B, C \in \mathcal{A}, A \cup B \cup C = X$ the order on $\mathcal{G}\,|A$ may depend on $g\,|B$ for a fixed $g\,|C$ but the way the order on $\mathcal{G}\,|A$ depends on $g\,|B$ may be independent of $g\,|C$. To be precise: The dependence of the order on $\mathcal{G}\,|A$ of $g\,|B$ is independent of $g\,|C$ if

$$\begin{array}{lcll} (y_1, y_2, y_3) & \succsim & (y_1', y_2, y_3) & \\ (y_1, y_2', y_3) & \precsim & (y_1', y_2', y_3) & \text{and} \\ (y_1, y_2, y_3') & \precsim & (y_1', y_2, y_3') & \text{imply} \\ (y_1, y_2', y_3') & \precsim & (y_1', y_2', y_3') & \end{array}$$

$$((y_1, y_1') \in \mathcal{G}\,|A\,, (y_2, y_2') \in \mathcal{G}\,|B\,, (y_{31}, y_3') \in \mathcal{G}\,|C)$$

If the change from y_2 to y_2' improves y_1' compared to y_1 for one value of y_3, we can not for some other value y_3' have the converse.

Conjecture 2 *If the dependence of the order on $\mathcal{G}\,|A$ of $g\,|B$ is independent of $g\,|C$ for all partitions A, B, C of X, then there exist a function*

$$f : \mathcal{G} \times \mathcal{B} \rightarrow \mathbb{R}$$

where $\mathcal{B}$ is the product algebra on $X \times X$. $f(\cdot, X \times X)$ is strictly monotonic in g and $f(g, \cdot)$ is additive.

Conjecture 3 *If in addition to the assumption from conjecture 2 $A \cup B$ contains at least three non null set for which assumption 3(b)iii holds for $\mathcal{B}$ a system of subsets of $A \cup B$, then the function f may chosen such that*

$$f(g, A_1 \times A_2) = 0 \text{ for } A_1 \times A_2 \in \mathcal{B} \text{ and } A_1 \cap A_2 = \emptyset.$$

If conjecture 2 is true we would have conditions with a reasonably nice economic interpretation implying for example existence of a utility function over time of the form

$$u(y_1, y_2, \cdots, y_t, \cdots) = \sum_{t=1}^{\infty} \sum_{s=1}^{\infty} u_{ts}(y_t, y_s)$$

If also conjecture 3 is true we may get conditions for a function of the form

$$u(y_1, y_2, \cdots, y_t, \cdots) = \sum_{t=1}^{\infty} u_t(y_t, y_{t+1})$$

9

Relations on function spaces

9.1 Introduction

The equivalence theorem (theorem 11, chapter 5) can now be proved by combining the results from chapters 5 and 8. The results in chapter 8 combined with the equivalence theorem yields representation results for first an independent subset of a function space and then for a relation on a function space. (Sections 9.2 and 9.3). As in chapter 8 several special cases are of interest. Section 9.4 covers the case where X is a finite set. $Y = \{0, 1\}$ gives subjective uncertainty in section 9.5. The minimal independence assumptions are given in section 9.6. And finally the consequences of special properties of the ranges of the functions for special representations are given in section 9.7.

9.2 Existence of $F : \mathcal{G} \to \mathbb{R}, f : \mathcal{G} \times \mathcal{A} \to \mathbb{R}$

Consider $((X, \mathcal{A}), Y, \mathcal{G}, Q)$, where X and Y are arbitrary sets and $\mathcal{A}$ a system of subsets of X. $\mathcal{G}$ is a space of functions defined on X with values in Y. Assuming that $\mathcal{G}$ is a mixture with respect to a finite partition $(A_i)_{i \in I}$ it can be regarded as a product with the notation from chapters 5 and 8

$$Q \subset \mathcal{G} = \prod_{i \in I} \mathcal{G} \,|A_i = \prod_{i \in I} S_i \supset Q$$

The definitions of essential, independence, total preorder on S_A, order topology on S_A, and the product topology on S can be repeated from section 5.2 page 40 and section 5.3 page 40.

Theorem 36 *Let $\mathcal{G}$ be a mixture with respect to an algebra $\mathcal{A}$ where $\mathcal{A}$ has more than three disjoint non-null sets. Let Q be independent, open and essential for all partitions $(A_i)_{i\in I}, (A_i \in \mathcal{A})$ of X and assume $\mathcal{G}\,|A_i$ connected. Then there exists functions*

$$F : \mathcal{G} \to \mathbb{R},\ f : \mathcal{G} \times \mathcal{A} \to \mathbb{R}$$
$$\textit{such that}$$
$$f(g, \cdot)\ \textit{ is additive, } F(g) = f(g, X)$$
$$\textit{and}$$
$$F(g) > 0 \Leftrightarrow g \in Q$$

Proof. $\mathcal{G}$ is a mixture with respect to $\mathcal{A}$, and Q is independent with respect to $\mathcal{A}$, so Q determines a commutative mean groupoid on $S_A = \mathcal{G}\,|A$ for any A with A^c non-null. Choose first an arbitrary $g^0 = (g_i^0)_{i\in I} \in \overline{Q} \setminus Q$. Choose then a representation F on $\mathcal{G}\,|A_i$ with $F(g_i^0) = 0$ for an arbitrary non-null set A_i. For all A for which $X \setminus (A_i \cup A)$ is non-null the mean groupoid representation can be uniquely extended. The extended representation will also be denoted F. F on all of $\mathcal{G}$ is then uniquely determined as

$$F(g) = \sum_{i\in I} F(g\,|A_i), F(g^0) = 0$$

for an arbitrary partition of X.

$f(g, A)$ is then defined as

$$f(g, A) = F(g \boxtimes_A g_0)$$

F obviously determines a total preorder on $\mathcal{G}$, and

$$Q = \{g \in \mathcal{G}\,|g \succ g^0\}$$

is proved using the definition of $\succsim$ first for any $\mathcal{G}\,|A$ for which A^c is not a null-set, and then on $\mathcal{G} = \mathcal{G}\,|A \times \mathcal{G}\,|A^c$ ∎

F also determines a midpoint operation $\circ$ on all of $\mathcal{G}$, in addition to the $\circ$ already determined on $\mathcal{G}\,|A$.

Definition 118 $(\mathcal{G}, \succsim, \circ, \Box)$ *where $\mathcal{G}$ is the function space, $\succsim$ the by Q induced total preorder on $\mathcal{G}$, $\circ$ the midpoint operation on $\mathcal{G}\,|A$ for all $A \in \mathcal{A}$, and $\Box \in \mathcal{G}/\sim$ with $s_0 \in \Box$ will be called a* mean groupoid with zero *determined by Q. $Q = \{g \in \mathcal{G}\,|g \succ \Box\}$.*

Remark 37 *Theorems 26 (page 95) and 36 shows that - under the assumptions of the theorems - the two structures $(\mathcal{G}, Q)$ and $(\mathcal{G}, \succsim, \square)$ are equivalent. Given $(\mathcal{G}, Q)$ one can find $(\mathcal{G}, \succsim, \square)$ and conversely. The two structures determine the same mean groupoid with a zero. $(\mathcal{G}, Q)$ can therefore also be given for example the order topology from $(\mathcal{G}, \succsim)$.*

9.3 Existence of $F : \mathcal{G} \times \mathcal{H} \to \mathbb{R}, f : \mathcal{G} \times \mathcal{H} \times \mathcal{A} \to \mathbb{R}$

X, Y and Z are arbitrary sets and $\mathcal{A}$ a system of subsets of X. $\mathcal{G}$ $(\mathcal{H})$ is a space of functions defined on X with values in $Y\,(Z)$. Finally $graph\mathcal{P} = Q \subset \mathcal{G} \times \mathcal{H}$. A reinterpretation of theorem 36 now gives a representation theorem for an independent relation $\mathcal{P}$ on $\mathcal{G} \times \mathcal{H}$.

Corollary 10 *Let $\mathcal{G} \times \mathcal{H}$ be a mixture with respect to $\mathcal{A}$ where $\mathcal{A}$ has more than three disjoint non-null sets. Let $Q = graph\mathcal{P}$ be independent, open and essential for all partitions $(A_i)_{i \in I}$, $(A_i \in \mathcal{A})$ of X and assume $(\mathcal{G} \times \mathcal{H})\,|A_i$ connected. Then there exists functions*

$$F : \mathcal{G} \times \mathcal{H} \to \mathbb{R}, \; f : \mathcal{G} \times \mathcal{H} \times \mathcal{A} \to \mathbb{R}$$
$$\textit{such that}$$
$$f(g, h, \cdot) \textit{ is additive, } F(g, h) = f(g, h, X)$$
$$\textit{and}$$
$$F(g, h) > 0 \Leftrightarrow (g, h) \in Q \Leftrightarrow g \in \mathcal{P}(h)$$

Proof. This is just theorem 36 with the translation $Y \to Y \times Z$ and $\mathcal{G} \to \mathcal{G} \times \mathcal{H}$ ■

Remark 38 $(\mathcal{G}, \mathcal{H}, \mathcal{P})$ *thus determines a mean groupoid with zero on $\mathcal{G} \times \mathcal{H}$, with $graph\mathcal{P}$ the positive elements from $\mathcal{G} \times \mathcal{H}$. The same remark applies with the obvious modifications to the following corollaries.*

Remark 39 *In particular $(\mathcal{G}, \mathcal{H}, \mathcal{P})$ thus determines a total preorder on $\mathcal{G} \times \mathcal{H}$ (and on $\mathcal{G} \times \mathcal{H}\,|A$, $A \in \mathcal{A}$.)*

$\mathcal{G} \times \mathcal{H} \subset (Y \times Z)^X$ can also be regarded as a space of functions defined on two copies of X - denoted $2X$ with values in $Y \cup Z$.

$$\mathcal{G} \times \mathcal{H} \subset (Y \times Z)^X = Y^X \times Z^X \subset (Y \cup Z)^{2X}$$

This means that the independence assumption can also be made with respect to X as a subset of $2X$.

Corollary 11 *Let $\mathcal{G} \times \mathcal{H}$ be a mixture with respect to $\mathcal{A}$ where $\mathcal{A}$ has more than two disjoint non-null sets. Let Q be independent, open and essential for all partitions $(A_i \cup A_j)_{i,j \in I}, (A_i, A_j) \in \mathcal{A}^2$ of $2X$ and assume $(\mathcal{G} \times \mathcal{H}) | A_i \cup A_j$ connected. Then there exist functions*

$$F_1 : \mathcal{G} \to \mathbb{R}, \; F_2 : \mathcal{H} \to \mathbb{R}$$
$$f_1 : \mathcal{G} \times \mathcal{A} \to \mathbb{R}, \; f_2 : \mathcal{H} \times \mathcal{A} \to \mathbb{R}$$

$$F_1(g) = f_1(g, X), F_2(g) = f_2(g, X)$$
$$f_1(g, \cdot) \text{ and } f_2(h, \cdot) \text{ are additive}$$
$$\text{and}$$
$$F_1(g) + F_2(h) > 0 \Leftrightarrow (g, h) \in Q \Leftrightarrow g \in \mathcal{P}(h)$$

Proof. This is just theorem 36 with the translation $Y \to Y \cup Z$ and $\mathcal{G} \to \mathcal{G} \times \mathcal{H}$ ∎

Assumption A page 43 will then imply $F_1 = -F_2$, and we get back theorem 26 as a special case.

9.3.1 $((X, \mathcal{A}), Y, \mathcal{G}, \mathcal{P})$ Existence of $F : \mathcal{G} \times \mathcal{G} \to \mathbb{R}, f : \mathcal{G} \times \mathcal{G} \times \mathcal{A} \to \mathbb{R}$

As another special case we have the, for applications, very important

Corollary 12 *Let $\mathcal{G}$ be a mixture with respect to $\mathcal{A}$ where $\mathcal{A}$ has more than three disjoint non-null sets. Let $\mathcal{P}$ be a relation on $\mathcal{G}$. Let $graph\mathcal{P} = Q$ be independent, open and essential for all partitions $(A_i)_{i \in I}, (A_i \in \mathcal{A})$ of X and assume $\mathcal{G} | A_i$ connected. Then there exists functions*

$$F : \mathcal{G} \times \mathcal{G} \to \mathbb{R}, \; f : \mathcal{G} \times \mathcal{G} \times \mathcal{A} \to \mathbb{R}$$
$$\text{such that}$$
$$f(g, h, \cdot) \text{ is additive, } F(g, h) = f(g, h, X)$$
$$\text{and}$$
$$F(g, h) > 0 \Leftrightarrow (g, h) \in Q \Leftrightarrow g \in \mathcal{P}(h)$$

Proof. This is just corollary 10 for $Y = Z$ and $\mathcal{G} = \mathcal{H}$ ∎

Remark 40 *$(\mathcal{G}, \mathcal{P})$ thus determines a mean groupoid with zero on $\mathcal{G} \times \mathcal{G}$, with graph$\mathcal{P}$ the positive elements from $\mathcal{G} \times \mathcal{G}$. The same remark applies with the obvious modifications to the following corollaries.*

Remark 41 *In particular $(\mathcal{G}, \mathcal{P})$ thus determines a total preorder on $\mathcal{G} \times \mathcal{G}$ (and on $\mathcal{G} \times \mathcal{G} \,|A$, $A \in \mathcal{A}$.)*

9.4 $X = \{1, 2, \cdots, n\}$ $(\prod_{i \in X} Y_i, \prod_{i \in X} Z_i, \mathcal{P})$

For $X = \{1, 2, \cdots, n\}$ the notation $(y_1, y_2, \cdots, y_n)$ is used for the values of the functions g. Theorem 36 reduces to

Corollary 13 *Let $Q \subset Y = (Y_1 \times Y_2 \times \cdots \times Y_n)$. Assume that $n > 3$, that $\# Y_i > 1$, that Q is essential, independent and open, and that Y_i is connected, then there exists a representation*

$$f : Y \to \mathbb{R}, w_i : Y_i \to \mathbb{R}, (i = 1, \cdots, n)$$
such that
$$f(y) = \sum w_i(y_i)$$
and
$$f(y) > 0 \Leftrightarrow y \in Q$$

Proof. This just theorem 36 for the special case considered ■

For $X = \{1, 2, \cdots, n\}$ the notation

$$(y_1, y_2, \cdots, y_n), (z_1, z_2, \cdots, z_n)$$

is used for the values of the functions g and h. Corollary 10 reduces to

Corollary 14 *Let $\mathcal{P}$ be a relation on*

$$(Y_1 \times Y_2 \times \cdots \times Y_n) \times (Z_1 \times Z_2 \times \cdots \times Z_n)$$

Assume that $n > 3$, that for $\forall i \in \{1, 2, \cdots, n\}, \# (Y_i \times Z_i) > 1$, and $(Y_i \times Z_i)$ is connected, that $\mathcal{P}$ is essential, independent and open, then there exists a representation

$$f : (Y_1 \times Y_2 \times \cdots \times Y_n) \times (Z_1 \times Z_2 \times \cdots \times Z_n) \to \mathbb{R}$$
such that
$$f(y_1, y_2, \cdots, y_n, z_1, z_2, \cdots, z_n) = \sum w_i(y_i, z_i)$$
and
$$f(y, z) > 0 \Leftrightarrow y \in \mathcal{P}(z)$$

Proof. This is just corollary 10 for the special case considered ■

Corollary 15 *Let $\mathcal{P}$ be a relation on $(Y_1 \times Y_2 \times \cdots \times Y_n)$. Assume that $n > 3$, that for $\forall i \in \{1, 2, \cdots, n\}, Y_i$ is connected, and $\#Y_i > 1$, and that $\mathcal{P}$ is essential, independent and open, then there exists a representation*

$$f(y_1, y_2, \cdots, y_n, y_1', y_2', \cdots, y_n') = \sum w_i(y_i, y_i')$$
and
$$f(y, y') > 0 \Leftrightarrow y \in \mathcal{P}(y')$$

Proof. This is just corollary 10 for the special case considered, for $Y = Z$. ■

Remark 42 *If assumption A (page 43) holds, $w_i(y_i, y_i) = 0$.*

9.5 $Y = Z = \{0, 1\}, (X, \mathcal{A}, \mathcal{P})$

This is the important special case where the relation is defined on $\mathcal{A}$, or equivalently on the class of indicator functions on X. The assumptions for a representation theorem are of course unchanged but it is useful to state the independence assumption.

Definition 119 (uncertainty space) *$(X, \mathcal{A}, \mathcal{P})$ is called an uncertainty space.*

Definition 120 (mixing) *The value of the function*

$$\boxtimes : \mathcal{A} \times \mathcal{A} \times \mathcal{A} \to \mathcal{A}$$
defined by
$$A \boxtimes_C B = (A \cap C) \cup (B \setminus C) \tag{mixing}$$

is called the mixing of A and B with respect to C.

Definition 121 (independent with respect to $\mathcal{A}$) *Let $(X, \mathcal{A}, \mathcal{P})$ be an uncertainty space. $\mathcal{P}$ is independent with respect to $\mathcal{A}$ if for $\forall A, A', B, B'$, and $C \in \mathcal{A}$*

$$\begin{matrix} A \in \mathcal{P}(A') & A \boxtimes_C B \notin \mathcal{P}(A' \boxtimes_C B') \\ B \in \mathcal{P}(B') & B \boxtimes_C A \notin \mathcal{P}(B' \boxtimes_C A') \end{matrix} \tag{NOT}$$

*can **not** hold.*

Theorem 37 *Let* $(X, \mathcal{A}, \mathcal{P})$ *be an uncertainty space, where* $(X, \mathcal{A})$ *has more than three disjoint subsets* A_i *for which* $A_i \in \mathcal{P}(\emptyset)$. *Let assumption* A *hold, let* $\mathcal{P}$ *be independent and open, and let* $\mathcal{A}\,|A_i$ *be connected. Then there exist additive functions*

$$\alpha : \mathcal{A} \to \mathbb{R}_+ \text{ and } \beta : \mathcal{A} \to \mathbb{R}_+$$
$$\text{where}$$
$$\alpha(A \backslash B) - \beta(B \backslash A) > 0 \Leftrightarrow A \in \mathcal{P}(B)$$

Proof. Assumption A implies that $f(0,0,A) = f(1,1,A) = 0$ and corollary 12 gives the result, by defining

$$\alpha(A) = f(1,0,A) \text{ and } \beta(A) = -f(0,1,A)$$

■

9.6 Minimal independence assumptions

Theorem 25 page 93 solved the problem of finding minimal independence assumptions on a preorder relation which would imply independence with respect to an algebra. This theorem also can be used to solve the same problem for a subset of product set or a subset of a function space

A definition will be used

Definition 122 (crossing) *Two sets* A, B *are crossing if* $A \cap B, A \setminus B, B \setminus A$, *and* $X \setminus (A \cup B)$ *are non-null.*

Theorem 38 *Let* $\mathcal{G}$ *be a connected mixture with respect to an algebra* $\mathcal{A}$. *If* $(\mathcal{G}, Q)$ *is independent with respect to crossing sets* $A, B \in \mathcal{A}$, *then* Q *is independent with respect to the algebra generated by* A, B.

Proof. When $(A \cup B)^c$ is non-null there is on $\mathcal{G}\,|(A \cup B)$ a total preorder which is independent with respect to $A \cap B, A \setminus B, A \setminus B$ and $A \triangle B$ (theorem 25). Independence with respect to $A^c \cap B^c$ then follows from independence with respect to $A \cup B$ because $A^c \cap B^c = (A \cup B)^c$ ■

9.7 $(Y_x, Q_x)_{x \in X}$

The proof of theorem 36 implies that a subset Q of a product space or of a function space $\mathcal{G}$ under the independence assumption implies (and is equivalent to) that $\mathcal{G}$ is totally preordered and a mean groupoid with a zero. $\mathcal{G}, \{(\mathcal{G} \times \mathcal{H})\}$ and $[\mathcal{G} \times \mathcal{G}]$ are commutative mean groupoids [with zero] and theorem 36, {corollary 10} and [corollary 12] give the representation theorems. The extra structure that with the independence assumption

$$\begin{array}{c} (Y_x, Q_x) \text{ where } (Y_x / \sim_x, \succsim_x, \circ_x, \Box_x) \\ (Y, Z, \mathcal{P}) \text{ where } \{((Y \times Z)_x / \sim_x, \succsim_x, \circ_x, \Box_x)\} \\ (Y_x, \mathcal{P}_x) \text{ where } [((Y_x \times Y_x) / \sim_x, \succsim_x, \circ_x, \Box_x)] \end{array}$$

are commutative mean groupoids (with or without zeros) give under consistency assumptions very special extra properties. Let the representations be

$$\begin{array}{ccc} g \in Q & \iff & F(g) = f(g, X) > 0 \\ g \succ h & \iff & F(g) = f(g, X) > F(h) = f(h, X) \\ (g_1, g_2) \succ (h_1, h_2) & \iff & \begin{array}{c} F(g_1, g_2) = f(g_1, g_2, X) > \\ F(h_1, h_2) = f(h_1, h_2, X) \end{array} \\ g \in \mathcal{P}(h) & \iff & F(g, h) = f(g, h, X) > 0 \\ (g_1, g_2) \in \mathcal{P}(h_1, h_2) & \iff & \begin{array}{c} F(g_1, g_2, h_1, h_2) = \\ f(g_1, g_2, h_1, h_2, X) > 0 \end{array} \\ y \in Q_x & \iff & F_x(y) > 0 \\ y \in \mathcal{P}_x(z) & \iff & F_x(y, z) > 0 \end{array}$$

In all the cases the results are trivial consequences of the isomorphism between a mean groupoid and (a subset of) the real numbers and the consistency assumptions which says that the natural order and midpoint structure on the function space derived from the structure on the range of the functions is preserved in the mean groupoid structure on the function space. The following lemma will be used repeatedly in the following subsections

Lemma 21 *If the representations above $F : \mathcal{G} \times \mathcal{H} \to \mathbb{R}$ and $f : \mathcal{G} \times \mathcal{H} \times \mathcal{A} \to \mathbb{R}$ are affine in $\mathcal{G} \times \mathcal{H}$, then*

$$\begin{array}{c} F = F_1 + F_2 \\ f = f_1 + f_2 \\ F_1 : \mathcal{G} \to \mathbb{R}, F_2 : \mathcal{H} \to \mathbb{R} \\ f_1 : \mathcal{G} \times \mathcal{A} \to \mathbb{R}, f_2 : \mathcal{H} \times \mathcal{A} \to \mathbb{R} \end{array}$$

If $\mathcal{G} = \mathcal{H}$ and assumption A holds then

$$F_1 = -F_2, f_1 = -f_2$$

Proof. Trivial ■

9.7.1 $(X, Y, \mathcal{G}, Q, (Q_x)_{x\in X})$

The general case is $(X, Y, \mathcal{G}, Q, (Q_x)_{x\in X})$ where $Q \subset \mathcal{G} \subset Y^X$ and $Q_x \subset Y$. The interpretation in this case is simply that Q is a subset of a set of functions $\mathcal{G}$ defined on $(X, \mathcal{A})$ with values in Y. Q_x is for each x a subset of $Y_x = \mathcal{G}(x) = \{g(x) | g \in \mathcal{G}\}$. For both structures the independence assumption implies that the representation theorem (theorem 36) will hold

Definition 123 (Consistency) *Consistency then means that*

$$\begin{aligned} g(x) \in Q_x, \forall x \in X &\implies g \in Q \\ g(x) \notin Q_x, \forall x \in X &\implies g \notin Q \end{aligned}$$

Theorem 39 *Consistency implies for the derived structures*

$$\succsim_x, \succsim, \circ_x \textit{and} \circ$$

$$\begin{array}{rcc} g(x) \succsim_x g'(x), \forall x \in X & \implies & g \succsim g' \\ g'(x) \circ_x g''(x) = g(x), \forall x \in X & \implies & g' \circ g'' = g \\ \textit{and for a mean groupoid with a zero} & & \\ g(x) \in \Box_x, \forall x \in X & \implies & g \in \Box \end{array}$$

Proof. Follows directly from the definitions ■

This means that we are back in the situation from section 8.9. Theorem 16 gives for $F(g) = \widetilde{F}(\widetilde{g}_x)_{x\in X} = \widetilde{F}(F_x(g(x)))_{x\in X} = \widetilde{F}(\widetilde{g})$

Corollary 16 *Assume consistency, then $\widetilde{F}$ and $\widetilde{f}(\cdot, A)$ are affine for all $A \in \mathcal{A}$ and $\widetilde{f}(\widetilde{g}, \cdot)$ is additive.*

Theorem 8.10 gives

Corollary 17 *Let $(Y_x, \succeq_x, \circ_x)$ have the representation $F_x : Y \to \mathbb{R}_x$. Assume that $\mathcal{G}$ with the structure derived from $(Y_x, \succeq_x, \circ_x)_{x\in X}$*

is measurable and consistent with the mean groupoid $(\mathcal{G}, \succsim, \circ)$ *assumed to be continuous, then there exists a normalization of* F *and* f *and measure* α *on* $\mathcal{A}$ *such that*

$$\begin{gathered} F(g \boxtimes_A g_0) = f(g, A) = \int_A F_x(g(x))\, d\alpha \\ g \in Q \Longleftrightarrow F(g) > 0 \\ g(x) \in Q_x \Longleftrightarrow F_x(g(x)) > 0 \end{gathered} \qquad (*)$$

The number of cases will be 18. A subset, a total preorder or a relation on $\mathcal{G}$ and on Y_x, all 9 cases with or without 0. In all the cases Y_x may itself be a function space mean groupoid. The two special cases

$$\begin{gathered} (\mathcal{G}, \succsim, \circ), (Y/\sim_x, \succeq_x, \circ_x) \\ (\mathcal{G}, \succsim, \circ, \square), (Y/\sim_x, \succeq_x, \circ_x, \square_x) \end{gathered}$$

with the representations

$$F(g \boxtimes_A g_0) = f(g, A) = \widetilde{f}(\widetilde{g}, A) = \int_A F_x(g(x))\, d\alpha$$

and

$$\begin{gathered} F(g \boxtimes_A g_0) = f(g, A) = \widetilde{f}(\widetilde{g}, A) = \\ \int_{X^+ \cap A} F_x(g(x))\, d\alpha + \int_{X^- \cap A} F_x(g(x))\, d\beta = \\ \int_A F_x^+(g(x))\, d\alpha + \int_A F_x^-(g(x))\, d\beta = \\ \int_A U(x, g(x))\, d\gamma \end{gathered}$$

were covered by theorems 17 and 31. The interpretations are that in the case where there is a total preorder both on the function space and on the range for each x, the representation of the total preorder on the function space is the expected value of the representations on the range.

Remark 43 *When the consistency assumption is only made separately for positive and negative* y *the representation can either be regarded as the expected value of a "kinked" representation of the mean groupoid for each* x *(the "kink" in* 0*), or as a representation with* (α, β) *uncertainty on* $\mathcal{A}$.

The special cases

$$\begin{gathered} ((\mathcal{G}, \succsim, \circ), ((Y_x, \mathcal{P}_x)_{x \in X})) = \\ ((\mathcal{G}, \succsim, \circ), (Y_x \times Y_x) / \sim_x, \succeq_x, \circ_x)_{x \in X}) \end{gathered}$$

$$\left((\mathcal{G},\mathcal{P}),(Y_x/\sim_x,\succsim_x,\circ_x)_{x\in X}\right)=$$
$$\left((\mathcal{G}\times\mathcal{G},\succsim,\circ),(Y_x/\sim_x,\succsim_x,\circ_x)_{x\in X}\right)$$

$$\left((\mathcal{G},\mathcal{P}),(Y_x/\sim_x,\succsim_x,\circ_x,\Box_x)_{x\in X}\right)=$$
$$\left((\mathcal{G}\times\mathcal{G},\succsim,\circ),(Y_x/\sim_x,\succsim_x,\circ_x,\Box_x)_{x\in X}\right)$$

$$\left((\mathcal{G},\mathcal{P}),((Y_x,\mathcal{P}_x)_{x\in X})\right)=$$
$$\left((\mathcal{G}\times\mathcal{G},\succsim,\circ),((Y_x\times Y_x)/\sim_x,\succsim_x,\circ_x,\Box_x)_{x\in X}\right)$$

will be of interest. Representations for other special cases can be obtained if needed.

9.7.2 $\left((\mathcal{G},\succsim,\circ),(Y_x,\mathcal{P}_x)_{x\in X}\right)=$
$\left((\mathcal{G},\succsim,\circ),(Y_x\times Y_x)/\sim_x,\succsim_x,\circ_x\right)$

The mean groupoid structure on the range is on $(Y_x\times Y_x)$, and the total preorder will have to be on $\mathcal{G}\times\mathcal{G}$. So this is the case where there is a total preorder $\succsim$ and a midpoint operation $\circ$ on $\mathcal{G}\times\mathcal{G}\subset(Y\times Y)^X$ and a relation $\mathcal{P}_x$ on Y_x. The definitions and results from chapter 8 can be directly applied.

Theorem 40 *Let*

$$(\mathcal{G}\times\mathcal{G},\succsim,\circ)$$
$$(Y_x\times Y_x/\sim_x,\succsim_x,\circ_x)$$

be commutative mean groupoids and

$$F:\mathcal{G}\times\mathcal{G}\to\mathbb{R},f:\mathcal{G}\times\mathcal{G}\times\mathcal{A}\to\mathbb{R}$$

be representations for $(\mathcal{G},\succsim,\circ)$. *And*

$$F_x:Y_x\times Y_x\to\mathbb{R}_x$$

a representation for $(Y_x\times Y_x/\sim_x,\succsim_x,\circ_x)$
Assume consistency, then the

$$\widetilde{F}:\widetilde{\mathcal{G}}\to\mathbb{R},\widetilde{f}:\widetilde{\mathcal{G}}\times\mathcal{A}\to\mathbb{R}$$
$$\text{defined by }\widetilde{g}(x)=F_x\left(g^1(x),g^2(x)\right)$$
$$F\left(g^1,g^2\right)=\widetilde{F}(\widetilde{g}),f(g,A)=f\left(g^1,g^2,A\right)=\widetilde{f}(\widetilde{g},A)$$

are such that $\widetilde{F}(\cdot)$ *and* $\widetilde{f}(\cdot,A)$ *are affine and* $\widetilde{f}(\widetilde{g},\cdot)$ *is additive*

Proof. Except for the notation this is theorem 16 ■

Definition 124 (measurable) *$\mathcal{G} \times \mathcal{G}$ with the structure derived from $((Y_x \times Y_x) / \sim_x, \succsim_x, \circ_x)_{x \in X}$ is measurable (with respect to $\mathcal{A}$ and for a given g_0, g_1) if $\widetilde{g} \in \widetilde{\mathcal{G}} \subset \mathbb{R}^X$ are Lebesgue measurable.*

Theorem 41 *Let $\mathcal{G}$ with the structure derived from*

$$((Y_x \times Y_x) / \sim_x, \succsim_x, \circ_x)_{x \in X}$$

(with the representation $F_x : Y \times Y \to \mathbb{R}$) be measurable and consistent with the mean groupoid $(\mathcal{G}, \succsim, \circ)$ assumed to be continuous, then there exists a normalization of F and f and measure α on $\mathcal{A}$ such that

$$\begin{array}{l} F(g \boxtimes_A g_0) = f(g, A) = f(g^1, g^2, A) = \widetilde{f}(\widetilde{g}, A) = \\ \int_A F_x(g(x))\, d\alpha = \int_A F_x(g^1(x), g^2(x))\, d\alpha \end{array} \qquad (**)$$

Proof. This is theorem 17 with changes in notation ■

Remark 44 *Remark 32 page 102 still applies.*

9.7.3 $(\mathcal{G}, \mathcal{P}), (Y_x / \sim_x, \succsim_x, \circ_x)$

Let $(\mathcal{G}, \mathcal{P})$ be a function space $\mathcal{G} \subset Y^X$ (where $(Y_x / \sim_x, \succsim_x, \circ_x)$ is a mean groupoid) with a relation $\mathcal{P}$ making $(\mathcal{G} \times \mathcal{G}, \succsim, \circ)$ a mean groupoid. $(Y_x / \sim_x, \circ_x)$ is a mean groupoid with the representation $F_x : Y_x \to \mathbb{R}_x$ and is isomorphic to $(\mathbb{R}_x, \geqq, \circ)$, so the definitions of $\widetilde{\mathcal{G}}$ and consistency can be extended to $(Y_x / \sim_x, \succsim_x, \circ_x) \times (Y_x / \sim_x, \succsim_x, \circ_x)$ and the definition of measurability from section 8.9 can be repeated.

Theorem 42 *Let*

$$(\mathcal{G} \times \mathcal{G}, \succsim, \circ)$$
and
$$(Y / \sim_x, \succsim_x, \circ_x)$$

be commutative mean groupoids and

$$F : \mathcal{G} \times \mathcal{G} \to \mathbb{R}, f : \mathcal{G} \times \mathcal{G} \times \mathcal{A} \to \mathbb{R}$$

be representations for $(\mathcal{G} \times \mathcal{G}, \succsim, \circ)$. Assume consistency, then the

$$\widetilde{F} : \widetilde{\mathcal{G}} \times \widetilde{\mathcal{G}} \to \mathbb{R}, \widetilde{f} : \widetilde{\mathcal{G}} \times \widetilde{\mathcal{G}} \times \mathcal{A} \to \mathbb{R}$$

defined by

$$\widetilde{g}(x) = F_x(g(x))$$
$$F(g) = \widetilde{F}\left(\widetilde{g^1}, \widetilde{g^2}\right), f\left(g^1, g^2, A\right) = \widetilde{f}\left(\widetilde{g^1}, \widetilde{g^2}, A\right)$$

are such that $\widetilde{F}(\cdot)$ *and* $\widetilde{f}(\cdot, A)$ *are affine and* $\widetilde{f}(\widetilde{g}, \cdot)$ *is additive .*

Proof. This is just an application of theorem 16 ■

Theorem 43

$$\widetilde{F}\left(\widetilde{g^1}, \widetilde{g^2}\right) = \widetilde{F_1}\left(\widetilde{g^1}\right) + \widetilde{F_2}\left(\widetilde{g^2}\right)$$
$$\widetilde{f}\left(\widetilde{g^1}, \widetilde{g^2}, A\right) = \widetilde{f_1}\left(\widetilde{g^1}, A\right) + \widetilde{f_2}\left(\widetilde{g^2}, A\right)$$

where $F_1, F_2, \widetilde{f_1}$ *and* $\widetilde{f_2}$ *are affine.*
If assumption A (page 43) holds

$$\widetilde{F}\left(\widetilde{g^1}, \widetilde{g^2}\right) = \widetilde{F_1}\left(\widetilde{g^1}\right) - \widetilde{F_1}\left(\widetilde{g^2}\right)$$
$$\widetilde{f}\left(\widetilde{g^1}, \widetilde{g^2}, A\right) = \widetilde{f_1}\left(\widetilde{g^1}, A\right) - \widetilde{f_1}\left(\widetilde{g^2}, A\right)$$

where F_1 *and* $\widetilde{f_1}(\cdot, A)$ *are affine*

$$F\left(\left(g^1, g^2\right) \boxtimes_A \left(g^1, g^2\right)_0\right) = f_1\left(g^1, A\right) - f_1\left(g^2, A\right) =$$
$$\widetilde{f_1}\left(\widetilde{g^1}, A\right) - \widetilde{f_1}\left(\widetilde{g^2}, A\right) = \int_A F_x\left(g^1(x)\right) d\alpha - \int_A F_x\left(g^2(x)\right) d\alpha =$$
$$\int_A \left(F_x\left(g^1(x)\right) - F_x\left(g^2(x)\right)\right) d\alpha$$

Proof. See lemma 21 and the proof of theorem 17 ■

Remark 45 *This is an important observation. The theorem means that consistency is only possible for this case if $\mathcal{P}$ is a total preorder, and there is nothing new covered by this theorem compared to section 8.9*

Remark 46 *The same observation hold for the case*

$$\left(\mathcal{G}, \mathcal{P}\right), \left(\left(Y_x, \mathcal{P}_x\right)_{x\in X}\right) = \left(\mathcal{G}, \mathcal{P}\right), \left(\left(Y_x \times Y_x\right) / \sim_x, \succsim_x, \circ_x\right)$$

so this case does not add anything compared to $\left(\left(\mathcal{G}, \succsim, \circ\right), \left(Y_x, \mathcal{P}_x\right)_{x\in X}\right)$

9.7.4 $(\mathcal{G}, \mathcal{P}), (Y_x / \sim_x, \succeq_x, \circ_x, \Box_x)$

Let $(\mathcal{G}, \mathcal{P})$ be a function space $\mathcal{G} \subset Y^X$ (where $(Y/\sim, \succeq_x, \circ_x)$ is a mean groupoid) with a relation $\mathcal{P}$ making $(\mathcal{G} \times \mathcal{G}, \succsim, \circ, \Box)$ a mean groupoid with zero, $(Y_x / \sim_x, \circ_x, \Box_x)$ is a mean groupoid and is isomorphic to $(\mathbb{R}_x, \geqq, \circ)$.

$$F_x : Y_x \to \mathbb{R}_x$$

is the homomorphism. The definitions of consistency and measurability from section 8.9 can be repeated. Consistency for a mean groupoid with zero again means separate consistency for positive and negative elements. It is also either assumed that

$$\left(g^1, g^1\right) \in \Box$$

or this relation is used to define $\Box$, the zero in $(\mathcal{G}, \succsim, \circ, \Box)$.

The only difference from subsection 9.7.3 is that the consistency assumption is made separately for positive and negative elements so we trivially get the representation

$$\begin{array}{c} \int_A \left(F_x\left(g^1(x)\right) - F_x\left(g^2(x)\right)\right)^+ d\alpha \\ + \int_A \left(F_x\left(g^1(x)\right) - F_x\left(g^2(x)\right)\right)^- d\beta \end{array} \qquad (***)$$

9.7.5 $\left((\mathcal{G}, \mathcal{P}), (Y_x, \mathcal{P}_x)_{x \in X}\right) =$
$((\mathcal{G} \times \mathcal{G}, \succsim, \circ, \Box), (Y \times Y) / \sim_x, \succeq_x, \circ_x, \Box_x)$

The case where the function space has a relation $\mathcal{P}$ on $\mathcal{G} \times \mathcal{G} \subset ((Y_x \times Y_x) \times (Y_x \times Y_x))_{x \in X}$, and the range a product $(Y_x \times Y_x) \times (Y_x \times Y_x)$ with a mean groupoid structure will be interesting in the later applications. The definitions and results from chapter 8 and subsection 9.7.4 can be directly applied.

Theorem 44 *Let*

$$(\mathcal{G} \times \mathcal{G} \times \mathcal{G} \times \mathcal{G}, \succsim, \circ, \Box)$$
$$\textit{and}$$
$$((Y_x \times Y_x) / \sim_x, \succeq_x, \circ_x, \Box_x)_{x \in X}$$

be commutative mean groupoids and

$$F : (\mathcal{G} \times \mathcal{G}) \times (\mathcal{G} \times \mathcal{G}) \to \mathbb{R}, f : (\mathcal{G} \times \mathcal{G}) \times (\mathcal{G} \times \mathcal{G}) \times \mathcal{A} \to \mathbb{R}$$

be representations for $(\mathcal{G}, \mathcal{P})$. Assume A, consistency and measurability, then the

$$\widetilde{F} : (\mathcal{G} \times \mathcal{G}) \times (\mathcal{G} \times \mathcal{G}) \to \mathbb{R}, \widetilde{f} : \widetilde{\mathcal{G}} \times \widetilde{\mathcal{G}} \times \mathcal{A} \to \mathbb{R}$$

defined by

$$F(g,h) = F\left((g^1, g^2), (h^1, h^2)\right) = \widetilde{F}\left(\widetilde{g}, \widetilde{h}\right)$$

$$f\left((g^1, g^2), (h^1, h^2), A\right) = \widetilde{f}\left(\widetilde{g}, \widetilde{h}, A\right)$$

are such that $\widetilde{F}(\cdot)$ and $\widetilde{f}(\cdot, A)$ are affine and $\widetilde{f}\left(\widetilde{g}, \widetilde{h}, \cdot\right)$ is additive and

$$\begin{array}{c} f\left((g^1, g^2), (h^1, h^2), A\right) = \\ \int_A \left(F_x\left(g^1(x), g^2(x)\right) - F_x\left(h^1(x), h^2(x)\right)\right)^+ d\alpha \\ + \int_A \left(F_x\left(g^1(x), g^2(x)\right) - F_x\left(h^1(x), h^2(x)\right)\right)^- d\beta \end{array} \qquad (\maltese)$$

Proof. Except for the notation this is representation $(***)$ ∎

The interpretation of this representation is that there for each x is a relation on Y_x making $Y_x \times Y_x$ a mean groupoid. The representation is

$$F_x : Y_x \times Y_x \to \mathbb{R}.$$

$F_x\left(g^1(x), g^2(x)\right)$ expresses how much $g^1(x)$ is preferred to $g^2(x)$, so

$$g^1(x) \in \mathcal{P}\left(g^2(x)\right) \iff F_x\left(g^1(x), g^2(x)\right) > 0$$

On $(\mathcal{G} \times \mathcal{G})$ there is the relation $\mathcal{P}$ making $(\mathcal{G} \times \mathcal{G}) \times (\mathcal{G} \times \mathcal{G})$ a mean groupoid with the representation

$$F : (\mathcal{G} \times \mathcal{G}) \times (\mathcal{G} \times \mathcal{G}) \to \mathbb{R}$$

$F\left((g^1, g^2), (h^1, h^2)\right)$ expresses how much g^1 is preferred to g^2 compared to how much h^1 is preferred to h^2. So

$$\left(g^1, g^2\right) \in \mathcal{P}\left(h^1, h^2\right) \iff F\left(\left(g^1, g^2\right), \left(h^1, h^2\right)\right) > 0$$

From the relation $\mathcal{P}$ on $(\mathcal{G} \times \mathcal{G})$ giving a total preorder on

$$(\mathcal{G} \times \mathcal{G}) \times (\mathcal{G} \times \mathcal{G})$$

we can define a relation $\mathcal{P}_0$ on $\mathcal{G}$

Definition 125 **($\mathcal{P}_0$)** *$g^1 \in \mathcal{P}_0(g^2)$ if $(g^1, g^2) \in \mathcal{P}(h^1, h^1)$*

From ✠ we then get

$$g^1 \in \mathcal{P}_0(g^2) \Longleftrightarrow$$
$$\int_X \left(F_x\left(g^1(x), g^2(x)\right)\right)^+ d\alpha + \int_X \left(F_x\left(g^1(x), g^2(x)\right)\right)^- d\beta > 0 \qquad (✠✠)$$

Remark 47 *$\mathcal{P}_0$ is defined from $\mathcal{P}$ (uniquely if assumption A and (✠) holds), but any $\mathcal{P}_0$ (given that (✠✠) holds) will also determine a unique $\mathcal{P}$ for which ✠ holds, so the two formulations are equivalent.*

Remark 48 *All the results in this section can also be given with just change in notation for the case where Y is itself a space of functions*

In chapter 18 (✠✠) will be a basic ingredient in a foundation for statistics where X will be a parameter space and Y random variables.

9.8 Notes

1. The results in this chapter come from [180] or are new.

2. Corollary 13 can be regarded as a foundation for $\mathbb{R}^n$ for $n > 3$. The two assumptions, open and connected, which uses terms using the independence assumption, can easily be replaced by assumptions not using defined concepts. An assumption implying $w_i(Y_i) = \mathbb{R}$ could also be invented so the number of axioms in this characterization of $\mathbb{R}^n$ for $n > 3$ is 5. The result is of course very different in form from the results of Hilbert, Tarski, and Tarski's students who have studied axiom systems for $\mathbb{R}^n$ for $n > 1$ and for other geometries in terms of points, lines, betweenness etc. (See for example Szmielev (1983) [172]). The idea that a midpoint operation is the natural first algebraic concept to be introduced can be found in both forms of axiom systems. In this book (and in [180]) the independence assumption (and the other assumptions) are used first to define the midpoint operation on Y and Y_i. With this midpoint operation Y_i is isomorphic to $(\mathbb{R}, >, \circ)$

and Y is isomorphic to $\mathbb{R}^n$, where Q is mapped into an open half-space.

3. Debreu (1960) [52] obtained some of the results from this book by using results from the theory of webs. Blaschke (1932 and 1934), Thomsen (1927 and 1929), Reidemeister (1929), Blaschke Bol (1938), and Chern (1935) [173, 174, 155, 28, 29, 30] and [37] generalized this theory, and it was further used developed in a series of papers Chern (1935 and 1936) Griffiths (1976 and 1977) Chern Griffiths (1978a,b,c) and Damiano (1983) [37, 38, 42, 40, 41, 89, 90] and [46]. The basic results in Blaschke Bol (1938) [30] and Debreu (1960) [52] do not use differentiability assumptions. The later theory starts with differentiable manifolds. The theorems in this book strongly suggest that this special structure is not needed for the basic linearity results.

4. It may be very interesting for some applications to observe that for example corollary 13 can be used on a preferred set $\mathcal{P}(y)$ for a fixed y. One would then get

$$y' \in \mathcal{P}(y) \iff f(y', y) = \sum w_i(y'_i, y) > 0$$

and as a special case - where the mean groupoid on the axis are the same for different y -

$$y' \in \mathcal{P}(y) \iff f(y', y) = \sum v_i(y) w_i(y'_i) > v_0(y)$$

After a transformation of y_i the preferred set is bounded by a hyperplane. For the special case one transformation of y_i can be used for all y and all preferred sets would be half-spaces. The hyperplanes would however not necessarily be parallel.

Part III

Relations on Measures

10

Relations on sets of probability measures

10.1 Introduction

If it is assumed that the domain X of the functions $\mathcal{G}, \mathcal{H}$ is not just a measurable space but a measure space, it becomes possible to assume that relations on $\mathcal{G} \times \mathcal{H}$ only depend on the distribution of (g, h). In this special case representation theorems can be obtained without using the machinery of the previous parts of the book. The conclusions of the theorems obtained are very similar to those from the earlier representation theorems, and some of the later integral representation theorems and decompositions can be used also on the representations from this chapter. These results are therefore included in this book even if the mathematics used is different - the Hahn-Banach theorem or just separating hyperplanes.

In section 10.3 the existence of preference functions and Bernoulli functions - i.e. order preserving real functions on a totally pre-ordered set of probability measures - is discussed. The representation theorems from chapter 2 can be used in many ways to get representations.

In section 10.4 the existence of real representation functions for preferences on spaces of measurable functions on a measure space are obtained under the assumption that only the distribution of the functions matters. Assuming convexity the representing function on the space of distributions is affine.

10.2 Definitions and mathematics

The concepts used in this chapter are

$$((X, \mathcal{A}, \alpha), (Y, \mathcal{Y}), (Z, \mathcal{Z}), \mathcal{G}, \mathcal{H}, \mathcal{P}, \Delta)$$

$(X, \mathcal{A}, \alpha)$ is a probability space with $\mathcal{A}$ an algebra of subsets of X, an arbitrary set, and $\alpha : \mathcal{A} \to [0,1]$ an additive normalized function $(\alpha(X) = 1)$.

$(Y,\mathcal{Y})$ and $(Z,\mathcal{Z})$ are arbitrary measurable spaces, and $\mathcal{G} \subset Y^X$, $\mathcal{H} \subset Z^X$ are sets of measurable functions, $\mathcal{P}$ is a relation on $\mathcal{G} \times \mathcal{H}$ with the interpretation $(g, h) \in graph\mathcal{P}$ or $g \in \mathcal{P}(h)$, if g is preferred to h, and σ is the coarsest product topology on $\mathcal{G} \times \mathcal{H}$ in which $graph\mathcal{P}$ is open.

Notation 6 *Denote by $\mathcal{Y} \otimes \mathcal{Z}$ the product $\sigma-$algebra.*

Definition 126 *Define $\mathcal{P}^{-1} : \mathcal{G} \to 2^{\mathcal{H}}$ by*

$$\mathcal{P}^{-1}(g) = \{h \in \mathcal{H} \,|\, g \in \mathcal{P}(h)\}$$

The classical case treated in von Neumann Morgenstern (1944) [141] is, that $\mathcal{G} = \mathcal{H}$ and that $\mathcal{P}$ is the $\succ$ relation from a total preorder $\succsim$.

The (joint) distribution of g, h is the function

$$\lambda : \mathcal{Y} \otimes \mathcal{Z} \to [0,1]$$
defined by
$$\lambda(B) = \alpha\{x \in X \,|\, (g(x), h(x)) \in B \in \mathcal{Y} \otimes \mathcal{Z}\}$$

Denote the mapping $(g, h) \mapsto \lambda$ by

$$\chi : \mathcal{G} \times \mathcal{H} \to \Delta = \{\lambda : \mathcal{Y} \otimes \mathcal{Z} \to [0,1] \,|\, \lambda \text{ a measure}\}$$
and let
$$\Lambda = \chi(\mathcal{G} \times \mathcal{H})$$

Denote by τ the finest topology on Λ in which $\chi : (\mathcal{G} \times \mathcal{H}, \sigma) \to (\Lambda, \tau)$ is continuous.

We shall later need the marginal distributions, so we define

$$\chi_1 : \mathcal{G} \to \Delta_1 = \{\lambda_1 : \mathcal{Y} \to [0,1] \,|\, \lambda_1 \text{ a measure}\}$$
by
$$\lambda_1(B) = \alpha(\{x \in X \,|\, g(x) \in B\}), \chi_1(\mathcal{G}) = \Lambda_1$$

and analogously for χ_2, λ_2, and Λ_2.

Denote by

$$\Gamma = \chi(gr\mathcal{P})$$

Γ is thus the set of distributions of pairs of stochastic variables, where the first variable is preferred to the second.

Definition 127 *$\mathcal{P}$ is state independent if*

$$\chi(g,h) = \chi(g',h') \Longrightarrow [g \in \mathcal{P}(h) \Leftrightarrow g' \in \mathcal{P}(h')].$$

Notation 7 (conv) *Denote by convA the convex hull of A.*

Definition 128 (convex) *$\mathcal{P}$ is convex, if*

$$conv\Gamma \cap conv\,(\Lambda \backslash \Gamma) = \emptyset.$$

Δ_1, Δ_2, and Δ are affine subsets of linear spaces. Denote by E the affine space spanned by Λ and by $\tilde{E}$ the linear space spanned by Λ.

10.3 Existence of a Bernoulli function

The existence of preference functions on Λ will not be discussed.

In the case where $\mathcal{G} = \mathcal{H}, \mathcal{P}$ can be a total preorder and the relation $\succsim$ can in the obvious way be defined on Δ_1

$$(graph\,(\succsim) \subset \Delta_1 \times \Delta_1).$$

Definition 129 (Bernouilli function) *A function*

$$f : \Delta_1 \to \mathbb{R}$$

is a Bernoulli function for $(\Delta_1, \succsim)$ if for $\delta, \delta' \in \Delta_1$

$$f(\delta) \geqq f(\delta') \Leftrightarrow \delta \succsim \delta'$$

Theorem 5 and Corollary 2 mean that if $\mathcal{P}$ is a total preorder and Δ_1 is connected and separable in the order topology or any finer topology then there will exist a Bernoulli function.

We shall as examples of the kind of assumptions which implies separability prove two theorems.

Theorem 45 *Let Y be a separable metric space, $\mathcal{Y}$ the Borel algebra and Δ_1 a connected totally preordered set of regular probability measures on $(Y, \mathcal{Y})$. If the order topology is coarser than $\sigma(rca\,(Y,\mathcal{Y}), C\,(Y))$, where $C\,(Y)$ is the set of real bounded continuous functions on Y, then there will exist a Bernoulli function $\sigma(rca\,(Y,\mathcal{Y}), C\,(Y))$-continuous on Δ_1.*

Proof. $\sigma(rca(Y,\mathcal{Y}), C(Y))$ on the set of probability measures is a separable metric topology, when Y is a separable metric space (see Billingsley (1968) [25] page 239, or Parthasarathy (1967) [144] Theorem 6.2). Δ_1 is a subset of a separable metric space and is therefore separable in $\sigma(rca(Y,\mathcal{Y}), C(Y))$, which is assumed to be finer than the order topology, so Δ_1 is a conneced, separable, totally preordered set and Corollary 2 gives the $\sigma(rca(Y,\mathcal{Y}), C(Y))$-continuous order homomorphism on Δ_1 ∎

Theorem 46 *Let $(Y,\mathcal{Y},\mu)$ be a measure space with null-sets $\mathcal{N}$, where $\mathcal{Y}\setminus\mathcal{N}$ is separable in the metric*

$$d(A,B)=\mu(A\triangle B)=\mu(A\setminus B\cup B\setminus A)$$

and let Δ_1 be a set of μ-continuous countably additive probability measures. If the norm topology is finer than the order topology for some total preorder in which Δ_1 is connected, then there exists a continuous Bernoulli function.

Proof. Let $(Y_1,Y_2,\cdots)$ be dense in $\mathcal{Y}\backslash\mathcal{N}$. The set of simple functions with rational values measurable with respect to the σ-algebra generated by $(Y_1,Y_2,\cdots)$ is countable and dense in $L_1(Y,\mathcal{Y},\mu)$. Δ_1 is metric and by the Radon-Nikodým theorem isomorphic to a subset of $L_1(Y,\mathcal{Y},\mu)$ and thus a separable and connected totally preordered set ∎

This theorem could be extended to Δ_1 a set of μ-continuous probability measures on an algebra using the generalized (Bochner) Radon Nikodým theorem (Dunford Schwartz (1958) [64] IV.9.14 page 315).

10.4 von Neumann Morgenstern preferences

Due to the difference in the mathematics involved the case where the ranges of the functions are finite, is treated first. The result is then obtained as a direct consequence of a separation theorem for convex sets in $\mathbb{R}^n$. The general case uses a basic separation theorem in linear spaces and yields a linear functional, which only under extra continuity assumptions can be represented as an integral.

10.4.1 The finite case

Let Y and Z be finite sets $Y = \{1, 2, \cdots, y\}$ and $Z = \{1, 2, \cdots, z\}$, let $\mathcal{G} \subset Y^X$, $\mathcal{H} \subset Z^X$ be sets of measurable functions

$$\begin{array}{c} g^{-1}(i) \in \mathcal{A} \text{ for } i \in Y \text{ for } g \in \mathcal{G} \\ h^{-1}(j) \in \mathcal{A} \text{ for } j \in Z \text{ for } h \in \mathcal{H} \end{array}$$

and let finally $\mathcal{P}$ be a relation on $\mathcal{G} \times \mathcal{H}$ with the interpretation $g \in \mathcal{P}(h)$, if g is preferred to h.

The (joint) distribution of (g, h) is the function $\lambda : Y \times Z \to [0, 1]$ defined by

$$\lambda(i, j) = \alpha(\{x \in X \,|g(x) = i, \, h(x) = j\})$$

We have trivially that

$$\lambda \in \Delta = \left\{ \lambda \in [0, 1]^{YZ} \left| \sum_i \sum_j \lambda(i, j) = 1 \right. \right\}$$

Denote the mapping $(g, h) \mapsto \lambda$ by $\chi : \mathcal{G} \times \mathcal{H} \to \Delta$ and let

$$\Lambda = \chi(\mathcal{G} \times \mathcal{H})$$

We shall later need the marginal distributions, so we define

$$\begin{array}{c} \chi_1 : \mathcal{G} \to \Delta_1 = \left\{ \lambda_1 \in [0, 1]^Y \,|\textstyle\sum_i \lambda_1(i) = 1 \right\} \\ \text{by} \\ \lambda_1(i) = \alpha(\{x \in X \,|g(x) = i\}), \chi_1(\mathcal{G}) = \Lambda_1 \end{array}$$

and analogously for χ_2 and λ_2.

$$\begin{array}{c} \chi_2 : \mathcal{H} \to \Delta_2 = \left\{ \lambda_2 \in [0, 1]^Z \left| \textstyle\sum_j \lambda_2(j) = 1 \right. \right\} \\ \text{by} \\ \lambda_2(j) = \alpha(\{x \in X \,|h(x) = j\}), \chi_2(\mathcal{H}) = \Lambda_2 \end{array}$$

Definition 130 (continuous (finite case)) *$\mathcal{P}$ is continuous, if*

$$conv\Gamma \text{ is open in } conv\Lambda$$

Theorem 47 (the finite case) *Let $\mathcal{P}$ be a state independent, convex, and continuous relation on $\mathcal{G} \times \mathcal{H} \subset (Y \times Z)^X$. Then there exists a function*

$$f : \mathcal{G} \times \mathcal{H} \to \mathbb{R} \text{ and a function } w : Y \times Z \to \mathbb{R}, \text{ such that}$$
$$f(g,h) = \sum w(i,j)\,\lambda(i,j) \text{ where } \lambda = \chi(g,h) \text{ and}$$
$$f(g,h) > 0 \iff g \in \mathcal{P}(h) \iff \chi(g,h) \in \Gamma$$

Proof. $conv\Gamma$ and $conv(\Lambda \setminus \Gamma)$ are disjoint (possibly empty) convex subsets of Δ. $conv\Gamma$ is open in $conv\Lambda$, so there exists a non-zero affine function $\tilde{f} : \Lambda \to \mathbb{R}$ such that $\tilde{f}(\Gamma) > 0$ and $\tilde{f}(\Lambda \setminus \Gamma) \leq 0$. $\tilde{f}$ can be extended to a linear function on $\mathbb{R}^{YZ}$ and therefore be represented by w. Define finally

$$f : \mathcal{G} \times \mathcal{H} \to \mathbb{R} \text{ by } f(g,h) = \tilde{f}(\chi(g,h))$$

■

Definition 131 (v. N. M. preference function) *The function $w : Y \times Z \to \mathbb{R}$ with the property*

$$g \in \mathcal{P}(h) \iff \sum w(i,j)\,\lambda(i,j) > 0$$

is called a von Neumann Morgenstern preference function.

The interpretation of $w(i,j)$ is "how much is i preferred to j", and g is then preferred to h if and only if the expected value of this preference for i over j is positive.

10.4.2 The general case

Definition 132 (internal point) *A point x in a subset A of a affine space E is an internal point in A if*

$$\text{for all } y \in E \text{ there exists an } \varepsilon > 0 \text{ such that}$$
$$ty + (1-t)x \in A \text{ for } t \leq \varepsilon$$

Definition 133 (continuous) *$\mathcal{P}$ is continuous if $conv\Gamma$ consists of internal points, and $conv(\Lambda \backslash \Gamma)$ contains at least one internal point.*

Lemma 22

$$conv\Gamma \text{ consists of internal points}$$
$$\Leftrightarrow$$
$$conv\chi\left(\mathcal{P}(h)\times\{h\}\right) \text{ and } conv\left(\{g\}\times\mathcal{P}^{-1}(g)\right)$$

consists of internal points for all g and h.

Proof. $\Rightarrow$ is trivial. Let $\chi(g',h') \in E$ and $\chi(g,h) \in conv\Gamma$. Then there exists $\varepsilon_1 > 0$ and $\varepsilon_2 > 0$ such that

$$t_1\chi(g',h) + (1-t_1)\chi(g,h) \in conv\chi\left(\mathcal{P}(h)\times\{h\}\right)$$
$$\text{and}$$
$$t_2\chi(g,h') + (1-t_2)\chi(g,h) \in conv\chi\left(\{g\}\times\mathcal{P}^{-1}(g)\right)$$

for $t_1 \le \varepsilon_1$ and $t_2 \le \varepsilon_2$. A convex combination of these two points is equal to

$$t\chi(g',h') + (1-t)\chi(g,h)$$
$$\text{and}$$
$$t\chi(g',h') + (1-t)\chi(g,h) \in conv\Gamma$$

for $t \le \min(\varepsilon_1,\varepsilon_2) > 0$ ■

Remark 49 *If a set consists of internal points, the convex hull of the set also consists of internal points.*

Theorem 48 (general case) *Let $\mathcal{P}$ be a state independent, convex, and continuous relation on $\mathcal{G}\times\mathcal{H} \subset (Y\times Z)^X$. Then there exists a function*

$$f:\mathcal{G}\times\mathcal{H}\to\mathbb{R} \text{ and an affine function}$$
$$\tilde{f}:\Lambda\to\mathbb{R}, \text{ such that}$$
$$f(g,h) = \tilde{f}(\chi(g,h)) > 0 \Leftrightarrow g\in\mathcal{P}(h)$$

Proof. $conv\Gamma$ and $conv(\Lambda\setminus\Gamma)$ are disjoint (possibly empty) convex subsets of Λ. $conv\Gamma$ have internal points, so by Theorem V.1.12 in Dunford Schwartz (1958) [64] page 412, there exists a non-zero affine function $\tilde{f}:\Lambda\to\mathbb{R}$ such that $\tilde{f}(\Gamma)\ge 0$ and $\tilde{f}(\Lambda\setminus\Gamma)\le 0$. It follows from the proof in [64] that $\tilde{f}$ is not constant on Γ and the existence of internal points in $conv(\Lambda\backslash\Gamma)$ and the assumption that $conv\Gamma$ consists of internal points implies that $\tilde{f}(\Gamma) > 0$. Define finally

$$f:\mathcal{G}\times\mathcal{H}\to\mathbb{R} \text{ by } f(g,h) = \tilde{f}(\chi(g,h))$$

■

10.4.3 Special cases

These main theorems have many special cases. The relation may only depend on the two marginal distributions. The case $\mathcal{G} = \mathcal{H}$ gives further possibilities for special cases, the relation may be transitive and/or total, or may be such that $g \in \overline{\mathcal{P}(g)}$ (local non-satiation) etc.

Definition 134 (v. N. M. preference function) *A function*

$$w : Y \times Z \to \mathbb{R} \text{ with the property}$$
$$f(g,h) = \tilde{f}(\chi(g,h)) = \int w(g(x), h(x))\, d\chi(g,h)$$

is called a von Neumann Morgenstern preference function.

Any affine function $\tilde{f}$ on Λ can of course be extended uniquely to a linear function – also denoted $\tilde{f}$ – on $\tilde{E}$, the smallest linear space containing Λ. If one wants the existence of a von Neumann Morgenstern preference function, i.e. an integral representation for $\tilde{f}$, continuity of the extended $\tilde{f}$ in some weak topology is necessary and sufficient. Continuity of $\tilde{f}$ on Λ in the topology τ will at least if Λ is rich enough follow automatically from the definition of τ. But only in special (finite dimensional) cases does continuity on $\tilde{E}$ follow from continuity on Λ, so an assumption that $graph\mathcal{P}$ is open in a weak topology for some choice of dual linear space $\tilde{E}'$ is not enough to get a representation theorem. An assumption that the extended $\tilde{f}$ **is** continuous is assuming the conclusion. So this important special case has no neat general characterization in terms of properties of the relation on the function space. If Λ is finite dimensional, if all measures have finite support, or if the preferences are determined by the first n moments of the distributions, then assuming that τ is coarser than the Euclidean topology will imply the existence of a von Neumann Morgenstern preference function.

Chapter 11 contains results which can be used to get the existence of a von Neumann Morgenstern preference function from Theorem 48.

To get to the classical case *von Neumann Morgenstern utility* (von Neumann Morgenstern (1944) [141] with a proof in the second edition 1947) we can first assume factor independence.

Definition 135 (factor independence) *$\mathcal{P}$ is factor independent if*

$$[\chi_1(g) = \chi_1(g') \text{ and } \chi_2(h) = \chi_2(h')] \\ \Rightarrow \\ [g \in \mathcal{P}(h) \Longleftrightarrow g' \in \mathcal{P}(h')]$$

Theorem 49 (finite case) *Let $\mathcal{P}$ be a state independent, factor independent, convex, and continuous relation on $\mathcal{G} \times \mathcal{H} \subset (Y \times Z)^X$. Then there exists $f_1 : \mathcal{G} \to \mathbb{R}$ and $f_2 : \mathcal{H} \to \mathbb{R}$ and $u_1 : Y \to \mathbb{R}$ and $u_2 : Z \to \mathbb{R}$ such that*

$$f_1(g) = \sum u_1(i)\lambda_1(i) \text{ and } f_2(h) = \sum u_2(j)\lambda_2(j) \\ g \in \mathcal{P}(h) \Longleftrightarrow f_1(g) > f_2(h)$$

Proof. Define

$$\sum_{12} : \mathbb{R}^{YZ} \to \mathbb{R}^{Y+Z} \text{ by } \sum_{12}(\lambda) = \left(\sum_j \lambda(\cdot, j), \sum_i \lambda(i, \cdot)\right)$$

factor independence implies that $\tilde{f} \circ \sum_{12}^{-1}$ is a real *function* defined on $\Lambda_1 \times \Lambda_2 \subset \mathbb{R}^{Y+Z}$. All functions are linear and the linear function $\tilde{f} \circ \sum_{12}^{-1} : \mathbb{R}^{Y+Z} \to \mathbb{R}$ can therefore be represented by f_1 and $-f_2$ and by u_1 and $-u_2$ ■

Similarly we get

Theorem 50 (general case) *Let $\mathcal{P}$ be a state independent, factor independent, convex, and continuous relation on $\mathcal{G} \times \mathcal{H} \subset (Y \times Z)^X$. Then there exists*

$$f_1 : \mathcal{G} \to \mathbb{R}, \ f_2 : \mathcal{H} \to \mathbb{R} \\ \tilde{f}_1 : \Delta_1 \to \mathbb{R} \text{ and } \tilde{f}_2 : \Delta_2 \to \mathbb{R}$$

where $\tilde{f}_1$ and $\tilde{f}_2$ are affine and

$$f_1(g) = \tilde{f}_1(\chi(g)) \text{ and } f_2(h) = \tilde{f}_2(\chi(h)) \text{ with} \\ g \in \mathcal{P}(h) \Longleftrightarrow f_1(g) > f_2(h)$$

Proof. The same proof as in the finite case can be used ■

Assume now in the finite case that $\mathcal{G} = \mathcal{H}$ and thus $Y = Z$. Then obviously u_1 and u_2 are real functions defined on the same sets, and it is trivial to show that for example an assumption of local non-satiation (i.e. $g \in \overline{\mathcal{P}(g)} \setminus \mathcal{P}(g)$) or an assumption that $\mathcal{P}$ is the $\succ$ relation from a total preorder implies that we have

Definition 136 (von Neumann Morgenstern utility) *i.e.*

$$\text{that } u_1 = u_2 = u$$

and that the preferences $\succsim$ *on* $\Lambda_1 = \Lambda_2$ *can be represented by*

$$\lambda_1 \succsim \lambda_2 \Longleftrightarrow \sum \lambda_1 (i) u (i) \geq \sum \lambda_2 (i) u (i)$$

Assume now in the general case that $\mathcal{G} = \mathcal{H}$ and thus $Y = Z$, and that we have von Neumann Morgenstern preferences, then obviously $w (y_1, y_2) = u_1 (y_1) - u_2 (y_2)$ where u_1 and u_2 are real functions defined on the same sets, and it is trivial to show that for example an assumption of local non-satiation (i.e. $g \in \overline{\mathcal{P} (g)} \setminus \mathcal{P} (g)$) or an assumption that $\mathcal{P}$ is the $\succ$ relation from a total preorder implies that we have

Definition 137 (von Neumann Morgenstern utility) *i.e.*

$$\text{that } u_1 = u_2 = u$$

and that the preferences $\succsim$ *on* $\Lambda_1 = \Lambda_2$ *can be represented by*

$$\lambda_1 \succsim \lambda_2 \Longleftrightarrow \int u (g (x)) \, d\lambda_1 \geq \int u (h (x)) \, d\lambda_2$$

Example 9 (the mean variance case) *In many applications it is assumed that preferences only depend on and are continuous in the first* k *moments of the distributions. In particular the case* $k = 2$ *is popular in applications in finance. Let* $\kappa : \Delta \to \mathbb{R}^5$ *denote the function giving the first two moments of a* $\lambda \in \Lambda$*. (Two means and three second moments).*

Definition 138 (moments determine $\mathcal{P}$) *The moments determine* $\mathcal{P}$ *if*

$$\kappa (\lambda (g', h')) = \kappa (\lambda (g, h)) \Rightarrow [g \in \mathcal{P} (h) \Leftrightarrow g' \in \mathcal{P} (h')]$$

Theorem 51 *Let* $\mathcal{P}$ *be a state independent, factor independent, convex, and continuous relation on* $\mathcal{G} \subset \mathbb{R}^X$*, where* $\mathcal{G}$ *is any class of random variables with first and second moments. Assume local non satiation and that the moments determine the preferences, then there exists a von Neumann Morgenstern preference function*

$$w (z_1, z_2) = (z_1 - \gamma_1 z_1^2) - (z_2 - \gamma_2 z_2^2) - (\gamma_2 - \gamma_1) z_1 z_2, \\ \text{with } \gamma_2 \geq \gamma_1 \geq 0$$

Proof. κ is linear, so $\tilde{f} \circ \kappa^{-1} : \mathbb{R}^5 \to \mathbb{R}$ is a linear function. Local non satiation implies $w(z_1, z_2) = 0$ for $z_1 = z_2$, and this again implies the properties of the coefficients in $w(z_1, z_2)$ ■

Example 10 *To illustrate the use of von Neumann Morgenstern preferences in this very special situation let n be the number of assets in a financial market. The only information needed to choose a portfolio is $b \in \mathbb{R}^n$ (the expected value of a unit of the assets) and $B \in \mathbb{R}^{n^2}$ (the second order moments, assumed to have an inverse). A portfolio $y \in \mathbb{R}^n$ is then preferred to another portfolio $x \in \mathbb{R}^n$ if and only if the expected value of the von Neumann Morgenstern preference function for the pair of random variables obtained from (y, x)*

$$W(y,x) = b(y-x) - \gamma_1 yBy + \gamma_2 xBx - (\gamma_2 - \gamma_1) xBy > 0.$$

For a given x W is maximized by

$$y = \frac{B^{-1}b}{2\gamma_1} - \frac{\gamma_2 - \gamma_1}{2\gamma_1} x$$

A fixed point of this equation gives the unique optimal portfolio

$$x = \frac{B^{-1}b}{\gamma_1 + \gamma_2}$$

The uniqueness of the optimal portfolio is of course exceptional. With positive uncertainty one would expect to get many optimal solutions.

Example to be continued as example 11 page 192.

$\mathcal{P}$ antisymmetric and $\mathcal{P}^c$ total

We can define uncertainty as $v : \mathcal{G} \times \mathcal{G} \to \mathbb{R}$ with

$$v(g,h) = v(h,g) = -\frac{1}{2}[f(g,h) + f(h,g)]$$

If the uncertainty is non-negative we can trivially not have both $g \in \mathcal{P}(h)$ and $h \in \mathcal{P}(g)$, so the relation is antisymmetric. If the uncertainty is 0, the relation $\mathcal{P}^c$ defined by $g \in \mathcal{P}^c(h) \Leftrightarrow g \notin \mathcal{P}(h)$ is total. This is the special case called regret theory. See Loomes

Sugden (1982) [128]. Results about implications of properties of the relation for the preference function should be easy to obtain at least if it is assumed that Λ contain "many" elements, for example all measures with finite support. For a von Neumann Morgenstern preference function similar results are equally trivial.

$\mathcal{P}$ transitive

Another special case is that $\mathcal{P}$ is transitive, this is by the definitions equivalent to

$$f(g,h) > 0 \text{ and } f(h,k) > 0 \Rightarrow f(g,k) > 0$$

If for a von Neumann Morgenstern preference function we have

$$w(i,k) \geq w(i,j) + w(j,k)$$

or the same condition in the general case, then $\mathcal{P}$ is transitive, and coversely if again Λ contains "many" elements.

10.5 Notes

1. Most of the new results in this chapter comes from [177] and [181].

2. Bernoulli (1738) [20] suggested that the St. Petersburg Paradox should be solved by maximizing expected utility

$$\int u(x)\,dx$$

instead of expected value

$$\int x dx$$

He proposed f defined by

$$f(p) = \int \log x dx$$

as a Bernoulli utility function

3. All assumptions in section 10.4 are the natural generalizations of the assumptions to get existence of von Neumann Morgenstern utility. See for example Herstein Milnor (1953) [95] or Blackwell Girshick (1954) [27]. State independence is often assumed implicitly by having preferences defined on distributions of stochastic variables or simply on sets of probability measures. The interpretation of the assumption is, that only the distribution of the random variables plays a role for preferences, but the joint distribution may be important, so the preferences are not determined by the two marginal distributions. To get von Neumann Morgenstern utility is assumed that the basic relation is total. With this assumption the convexity assumption is also a consequence of usual assumptions. The interpretation of the assumption is that if one stochastic variable g is preferred to another g', and h is preferred to h', then a random variable where one gets g with probability $\frac{1}{2}$ and h with probability $\frac{1}{2}$ will also be preferred to g' with probability $\frac{1}{2}$ and h' with probability $\frac{1}{2}$, assuming that the distributions of these random variables are in Λ. A variant of the convexity assumption is that

$$\lambda_{1n} \succsim \lambda_{2n} \text{ for all } n \Rightarrow \sum \alpha_n \lambda_{1n} \succsim \sum \alpha_n \lambda_{2n}$$
$$\left(\sum \alpha_n = 1, \alpha_n \geq 0 \ \forall n\right)$$

and if in addition there exists an n such that

$$\lambda_{1n} \succ \lambda_{2n} \text{ then } \sum \alpha_n \lambda_{1n} \succ \sum \alpha_n \lambda_{2n}.$$

Axiom H1 in Blackwell Girshick (1954) [27] (page 105).

The relations $\succ$ and $\succsim$ are transitive, and it follows then easily that

$$\lambda_1 \succ \mu_1 \text{ and } \lambda_2 \succ \mu_2 \Rightarrow \alpha\lambda_1 + (1-\alpha)\lambda_2 \succ \alpha\mu_1 + (1-\alpha)\mu_2 \text{ and}$$
$$\lambda_1 \succsim \mu_1 \text{ and } \lambda_2 \succsim \mu_2 \Rightarrow \alpha\lambda_1 + (1-\alpha)\lambda_2 \succsim \alpha\mu_1 + (1-\alpha)\mu_2$$

so $gr \succ$ and $(gr \succ)^c$ are both convex, and the convexity assumption made here is in the special case, a total preorder, the usual.

The usual continuity assumption is

$$\{\alpha \in [0,1] \,|\alpha\lambda_1 + (1-\alpha)\lambda_2 \succ \lambda\} \text{ and}$$
$$\{\alpha \in [0,1] \,|\alpha\lambda_1 + (1-\alpha)\lambda_2 \prec \lambda\ \} \text{ are open}$$

(For example in the form of assumption H2 in Blackwell Girshick (1954) [27] page 106: If $\lambda_2 \succ \lambda_2 \succ \lambda_3$ then there exists α and β such that $\alpha\lambda_1 + (1-\alpha)\,\lambda_3 \succ \lambda_2$ and $\beta\lambda_1 + (1-\beta)\,\lambda_3 \prec \lambda_2$.) This assumption is a consequence of continuity, and is for the finite case in fact weaker. In the general case convexity is needed to give the equivalence between the two concepts of continuity, but assuming convexity the two continuity properties are trivially equivalent. (See Lemma 22 page 139).

Part IV

Integral Representations

11

A general integral representation by Birgit Grodal

11.1 Introduction

1. Chapter 8 studied a totally preordered set $(\mathcal{G}, \succsim)$ of functions $g : X \to Y$, where $(X, \mathcal{A})$ is a measurable set and Y an arbitrary set. $\mathcal{G}$ was under an independence condition shown to be a commutative mean groupoid $(\mathcal{G}, \succsim, \circ)$. This implied (theorem 26 page 95) the existence of functions

$$F : \mathcal{G} \to \mathbb{R},\ f : \mathcal{G} \times \mathcal{A} \to \mathbb{R}$$
with
$$f(g, A) = F(g \boxtimes_A g_0) \text{ and } F(g) = f(g, X)$$
with $f(g, \cdot)$ additive in $\mathcal{A}$
such that
$$g \succsim h \Leftrightarrow F(g) \geqq F(h)$$

($\mathcal{A}$ only assumed to be an algebra).

2. Chapter 9 studied first (section 9.2) $((X, \mathcal{A}), Y, \mathcal{G}, Q)$, where X and Y are arbitrary sets and $\mathcal{A}$ a system of subsets of X. $\mathcal{G}$ is a space of functions defined on X with values in Y, and $Q \subset \mathcal{G}$. Again under an independence condition will there exists functions (see corollary 10 page 115)

$$F : \mathcal{G} \to \mathbb{R},\ f : \mathcal{G} \times \mathcal{A} \to \mathbb{R}$$
such that
$$f(g, \cdot) \text{ is additive, } F(g) = f(g, X)$$
and
$$F(g) > 0 \Leftrightarrow g \in Q$$

3. Chapter 9 then (section 9.3) studied $((X, \mathcal{A}), Y, Z, \mathcal{G}, \mathcal{H}, \mathcal{P})$ where X, Y and Z are arbitrary sets and $\mathcal{A}$ a system of

subsets of X. $\mathcal{G}$ $(\mathcal{H})$ is a space of functions defined on X with values in $Y\,(Z)$, and $\mathcal{P}$ a relation on $\mathcal{G} \times \mathcal{H}$. Finally $graph\mathcal{P} = Q \subset \mathcal{G}\times\mathcal{H}$. An application (corollary 12 page 116) of the previous result, then gives the existence of functions

$$F : \mathcal{G} \times \mathcal{H} \to \mathbb{R},\ f : \mathcal{G} \times \mathcal{H} \times \mathcal{A} \to \mathbb{R}$$
such that
$$f\,(g, h, \cdot) \text{ is additive, } F\,(g, h) = f\,(g, h, X)$$
and
$$F\,(g, h) > 0 \Leftrightarrow (g, h) \in Q \Leftrightarrow g \in \mathcal{P}\,(h)$$

4. As a further special case (section 9.3.1) ($Y = Z$ and $\mathcal{G} = \mathcal{H}$ $((X, \mathcal{A}), Y, \mathcal{G}, \mathcal{P})$) we obtained (corollary 15) the existence of functions

$$F : \mathcal{G} \times \mathcal{G} \to \mathbb{R},\ f : \mathcal{G} \times \mathcal{G} \times \mathcal{A} \to \mathbb{R}$$
such that
$$f\,(g, h, \cdot) \text{ is additive, } F\,(g, h) = f\,(g, h, X)$$
and
$$F\,(g, h) > 0 \Leftrightarrow (g, h) \in Q \Leftrightarrow g \in \mathcal{P}\,(h)$$

5. Let $(X, \mathcal{A}, \alpha)$ is a probability space with $\mathcal{A}$ an algebra of subsets of X, an arbitrary set, and $\alpha : \mathcal{A} \to [0, 1]$ an additive normalized function $(\alpha\,(X) = 1)$. $(Y,\mathcal{Y})$ and $(Z,\mathcal{Z})$ are arbitrary measurable spaces, and $\mathcal{G} \subset Y^X$, $\mathcal{H} \subset Z^X$ are sets of measurable functions, $\mathcal{P}$ is a relation on $\mathcal{G} \times \mathcal{H}$ with the interpretation ($(g, h) \in graph\mathcal{P}$ or $g \in \mathcal{P}\,(h)$, if g is preferred to h). Let $\mathcal{P}$ be a relation on $\mathcal{G} \times \mathcal{H}$. Assuming that $\mathcal{P}$ only depends on the joint distribution $(\in \Lambda)$ we obtained (in chapter 10, theorem 48, page 139) a function

$$f : \mathcal{G} \times \mathcal{H} \to \mathbb{R} \text{ and an affine function}$$
$$\tilde{f} : \Lambda \to \mathbb{R}, \text{ such that}$$
$$f\,(g, h) = \tilde{f}\,(\chi\,(g, h)) > 0 \Leftrightarrow g \in \mathcal{P}\,(h)$$

In all of these representations with additive or affine functions it is very important to obtain integral representations. Fortunately the first four representations are special cases of each other and the proofs will be carried out for case 1. The next three cases will then be easy corollaries.

1. In this chapter we obtain the existence of a measurable function

$$u : X \times Y \to \mathbb{R}$$
and a measure
$$\mu : \mathcal{A} \to \mathbb{R}_+$$
such that
$$f(g, A) = \int_A u(x, g(x))\, d\mu$$
so
$$\int_X u(x, g(x))\, d\mu > \int_X u(x, h(x))\, d\mu \Leftrightarrow g \succsim h$$

(Theorems 52, 54, and 53).

2. In this chapter we obtain the existence of a measurable function

$$u : X \times Y \to \mathbb{R}$$
and a measure
$$\mu : \mathcal{A} \to \mathbb{R}_+$$
such that
$$f(g, A) = \int_A u(x, g(x))\, d\mu$$
so
$$\int_X u(x, g(x))\, d\mu > 0 \Leftrightarrow g \in Q$$

(corollary 18).

3. In this chapter is obtained a function

$$u : X \times Y \times Z \to \mathbb{R}$$
and a measure
$$\mu : \mathcal{A} \to \mathbb{R}$$
such that
$$f(g, h, A) = \int_X u(x, g(x), h(x)) d\mu$$
so
$$\int_X u(x, g(x), h(x)) d\mu > 0 \Leftrightarrow (g, h) \in Q \Leftrightarrow g \in \mathcal{P}(h)$$

(corollary 19).

4. In this chapter is obtained a function

$$u : X \times Y \times Y \to \mathbb{R}$$

and a measure

$$\mu : \mathcal{A} \to \mathbb{R}$$

such that

$$f(g,h,A) = \int_A u(x, g(x), h(x))d\mu$$

so

$$\int_X u(x, g(x), h(x))d\mu > 0 \Leftrightarrow (g,h) \in Q \Leftrightarrow g \in \mathcal{P}(h)$$

(corollary 20).

5. Will not be covered here.

For the basic concepts used in this and the following chapter the reader is referred to Halmos (1961) [93] and Bourbaki (1939 -) [33].

11.2 Existence of $u : X \times Y \to \mathbb{R}$ with $f(g,A) = \int_A u(x, g(x))d\mu$

Let $(X, \mathcal{A}, \mu)$ be a positive σ-finite measure space, and (Y, d) a separable metric space. $\mathcal{G}$ is a set of measurable functions defined on X with values in Y, and $\succsim$ is a total preorder on $\mathcal{G}$.

Let $\mathcal{U}$ be the finest uniformity on Y compatible with the topology on Y and let t be the topology of $\mathcal{U}$-uniform convergence on $\mathcal{G}$.

Definition 139 (weakly (strongly) quasi-continuous) *We say that a measurable function $u : X \times Y \to \mathbb{R}$ is weakly [strongly] quasi-continuous if for every sequence (g_n) in $\mathcal{G}$ and $g \in \mathcal{G}$ where $g_n \to g$ a.e., we have that*

$$u(\cdot, g_n(\cdot)) \to u(\cdot, g(\cdot)) \text{ a.e. } [\text{ in } L_1(X, \mathcal{A}, \mu)].$$

In the following lemma we shall assume that we have a function $f : \mathcal{G} \times \mathcal{A} \to \mathbb{R}$ such that $f(\cdot, A)$ is strictly monotonic on $\mathcal{G}|A$ for all $A \in \mathcal{A}$, and such that for all $g \in \mathcal{G}, f(g, \cdot) : \mathcal{A} \to \mathbb{R}$ is a countable additive measure on $\mathcal{A}$, which is continuous with respect to μ. Conditions on $(\mathcal{G}, \succsim)$ implying existence of such a function is given in theorem 26. Also we let $(y_n)_{n=1}^{\infty}$ be a dense subset in

Y, and denote the corresponding constant functions from X to Y by $(\overline{y}_n)_{n=1}^{\infty}$. Moreover we let, for all $g \in \mathcal{G}, u_g : X \to Y$, denote an arbitrary but fixed, Radon-Nikodým derivative of the measure $f(g, \cdot)$ with respect to μ. With this notation we have the following lemma.

Lemma 23 *If $(\mathcal{G}, \succsim)$ is an independent mixture containing the constant functions, and $f(\cdot, X)$ is continuous on $(\mathcal{G}, t)$ then for all $g \in \mathcal{G}$*

$$u_g(x) = \lim_{y_n \to g(x)} u_{\overline{y}_n}(x) \text{ a.e. } x \in X$$

Proof. Let $g \in \mathcal{G}$. We will first prove that there exists a metric d_g on Y compatible with the topology and such that $f(\cdot, X)$ is continuous at g with respect to d_g-uniform convergence.

Indeed consider the family $(d_i)_{i \in I}$ of all metrics on Y, such that d_i is continuous on $Y \times Y$. Then the family $(U_i)_{i \in I}$, where

$$U_i = \{(y, z) \in Y \times Y | d_i(y, z) < 1\}$$

is a base for the uniformity $\mathcal{U}$. (See Bourbaki (1939-) [33] chap. 9, Sect. 1, ex. 5 (page 215 in the English edition), where we moreover have used that Y is a metric space). We therefore have that $(V_i)_{i \in I}$, where

$$V_i = \{h \in \mathcal{G} | d_i(h(x), g(x)) < 1 \text{ for all } x \in X\}$$

is a base for the neighborhoods of g in $(\mathcal{G}, t)$. Since $f(\cdot, X)$ is continuous there therefore exists a decreasing sequence (V_{i_n}) such that $|f(g, X) - f(h, X)| < \frac{1}{n}$ for all $h \in V_{i_n}$. Now let (d_{i_n}) be the corresponding sequence of metrics and define the metric d_g on Y by $d_g = \sum_{k=1}^{\infty} \frac{1}{2^k} \min(d_{i_k}, 1)$. d_g has the wanted property, since $|f(h, X) - f(g, X)| < \frac{1}{n}$ for all $h \in \mathcal{G}$ such that $d_g(h(x), g(x)) < \frac{1}{2^n}$ for all x.

To prove the lemma we shall prove that

$$u_g(x) \geq \limsup_{y_n \to g(x)} u_{\overline{y}_n}(x) \text{ for a.e. } x \in X$$

It then follows by symmetry that

$$u_g(x) \leq \liminf_{y_n \to g(x)} u_{\overline{y}_n}(x) \text{ for a.e. } x \in X,$$

and we therefore have

$$\underset{y_n \to g(x)}{limsup}\, u_{\overline{y}_n}(x) = u_g(x) \text{ for a.e. } x \in X.$$

In order to prove that $u_g(x) \geq \underset{y_n \to g(x)}{limsup} u_{\overline{y}_n}(x)$, define for each $\varepsilon > 0$,

$$S_\varepsilon = \{x \in X | u_g(x) + \varepsilon < \underset{y_n \to g(x)}{limsup} u_{\overline{y}_n}(x)\}.$$

We shall prove that $\mu(S_\varepsilon) = 0$ for all $\varepsilon > 0$.

First we remark that $S_\varepsilon \in \mathcal{A}$. Indeed let the functions (t_n) be defined by

$$t_n(x) = \inf\{k | d_g(y_k, g(x)) < \frac{1}{2^n} \text{and } u_{\overline{y}_k}(x) > u_g(x) + \varepsilon\}$$

Since the functions (t_n) are measurable and

$$S_\varepsilon = \{x \in X | \forall n : t_n(x) < +\infty\}$$

S_ε is measurable.

Now assume that there for some $\varepsilon > 0$ exists a measurable subset S'_ε of S_ε such that $0 < \mu(S'_\varepsilon) < +\infty$. We shall obtain a contradiction by constructing functions g_n, mixing finitely many of the constant functions and the function g, such that

$$d_g(g_n(x), g(x)) < \tfrac{1}{2^n} \text{ for all } x \in X$$
and such that
$$u_{g_n}(x) \geq u_g(x) \text{ for all } x \in X$$

and on a fixed set of positive measure S''_ε,

$$u_{g_n}(x) > u_g(x) + \varepsilon \tag{11.1}$$

This clearly gives a contradiction for n sufficiently large. Indeed

$$d_g(g_n(x), g(x)) < \tfrac{1}{2^n} \text{ for all } x \Longrightarrow$$
$$|f(g_n, X) - f(g, X)| < \tfrac{1}{n}$$
but
$$u_{g_n}(x) \geq u_g(x) \text{ and } u_{g_n}(x) > u_g(x) + \varepsilon \text{ for } x \in S''_\varepsilon \Longrightarrow$$
$$f(g_n, X) = \int_X u_{g_n}(x) d\mu \geq$$
$$\int_X u_g(x) d\mu + \varepsilon\mu(S''_\varepsilon) = f(g, X) + \varepsilon\mu(S''_\varepsilon) \text{ for all } n$$

We first construct S''_ε. Define for each n,

$$k_n = \inf\{k | \mu(S'_\varepsilon \cap \{x \in X | t_n(x) > k\}) \leq \frac{1}{3^n}\mu(S'_\varepsilon)\}$$

and define

$$S''_\varepsilon = S'_\varepsilon \cap \{x \in X | t_n(x) \leq k_n \text{ for all } n\}$$

Then $S''_\varepsilon \in \mathcal{A}$, $\mu(S''_\varepsilon) > 0$. Moreover each function t_n has at most finitely many values on S''_ε.

We now define the functions (g_n), $g_n : X \to Y$ by

$$g_n(x) = \begin{cases} y_{t_n}(x) & \text{for } x \in S''_\varepsilon \\ g(x) & \text{for } x \notin S''_\varepsilon \end{cases}$$

Since $\mathcal{G}$ is a mixture containing the constant functions, $g_n \in \mathcal{G}$ for all n. By the definition of the functions (t_n), we also have that $d_g(g_n(x), g(x)) < \frac{1}{2^n}$ for all $x \in X$. Moreover a.e. on S''_ε, $u_{g_n}(x) = u_{\overline{y}_{tn}}(x)$ and therefore, also by the definition of the functions (t_n)

$$u_{g_n}(x) > u_g(x) + \varepsilon \text{ for a.e. } x \in S''_\varepsilon.$$

Since

$$u_{g_n}(x) = u_g(x) \text{ for a.e. } x \notin S''_\varepsilon$$

the functions (g_n) have the properties which were shown to lead to a contradiction.

Therefore for all $\varepsilon > 0, \mu(S_\varepsilon) = 0$, which was to be proved ■

Theorem 52 *Let $(\mathcal{G}, \succsim)$ be an independent and connected mixture with respect to the σ-algebra $\mathcal{A}$ containing more than two non-null sets and containing the constant functions. Assume that $(\mathcal{G}, \succsim)$ is continuous with respect to the measure μ and that the order topology is coarser than t.*

Then there exists a weakly quasi-continuous measurable function $u : X \times Y \to \mathbb{R}$ such that, when the function $f : \mathcal{G} \times \mathcal{A} \to \mathbb{R}$ is defined by

$$f(g, A) = \int_A u(x, g(x))d\mu,$$

then $f(\cdot, A)$ is strictly monotonic on $\mathcal{G}|A$ for all $A \in \mathcal{A}$.

Proof. From theorem 26 we know that there exists a function $f : \mathcal{G} \times \mathcal{A} \to \mathbb{R}$ such that $f(\cdot, A)$ is strictly monotonic on $\mathcal{G}|A$ and $f(g, \cdot)$ is countable additive and continuous with respect to μ.

We will prove the existence of a measurable function $u : X \times Y \to \mathbb{R}$, such that for all $A \in \mathcal{A}$ and all

$$g \in \mathcal{G}, f(g, A) = \int_A u(x, g(x)) d\mu.$$

We first remark that a measurable function $u : X \times Y \to \mathbb{R}$ has the property above if and only if for all $g \in \mathcal{G}, u(\cdot, g(\cdot))$ is a Radon-Nikodým derivative of $f(g, \cdot)$. In order to construct a measurable function u with this property we utilize the Radon-Nikodým derivative of the functions $f(\overline{y}_n, \cdot)$ corresponding to the countable many constant functions $(\overline{y}_n)_{n=1}^{\infty}$.

Indeed define the function $u : X \times Y \to \mathbb{R}$ by

$$u(x,y) = \begin{cases} \underset{y_n \to y}{limsup}\, u_{\overline{y}_n}(x) & \text{if } \underset{y_n \to y}{limsup}\, u_{\overline{y}_n}(x) < +\infty \\ 0 & \text{if } \underset{y_n \to y}{limsup}\, u_{\overline{y}_n}(x) = +\infty \end{cases}$$

We first prove that u is measurable. Consider the function $\widetilde{u}$, where

$$\widetilde{u}(x,y) = \underset{y_n \to y}{limsup}\, u_{\overline{y}_n}(x).$$

For all $(x, y) \in X \times Y$ we have that

$$\widetilde{u}(x,y) = \lim u^n(x,y)$$

where the functions u^n are defined by

$$u^n(x,y) = \sup\{u_{\overline{y}_k}(x) | d(y_k, y) < \frac{1}{n}\}$$

However the functions (u^n) are measurable, since for every λ,

$$\{(x,y) \in X \times Y | u^n(x,y) > \lambda\} =$$
$$\{(x,y) \in X \times Y | \exists k : d(y_k, y) < \tfrac{1}{n} \text{ and } u_{\overline{y}_k}(x) > \lambda\} =$$
$$\bigcup_{k=1}^{\infty} (\{x \in X | u_{\overline{y}_k}(x) > \lambda\} \times \{y | d(y_k, y) < \tfrac{1}{n}\})$$

is measurable. Therefore $\widetilde{u}$ is measurable and consequently also the function u.

We moreover have to show that for all $g \in \mathcal{G}$,

$$u_g(x) = u(x, g(x)) \text{ for a.e. } x \in X$$

where u_g is a Radon-Nikodým derivative of $f(g, \cdot)$. To do this we first remark that the function $f(\cdot, X)$ is continuous on $(\mathcal{G}, t)$. Indeed $f(\cdot, X)$ is strictly monotonic and we have assumed that the order topology is coarser than t. Therefore according to lemma 23 for any $g \in \mathcal{G}$,

$$u_g(x) = \lim_{y_n \to g(x)} u_{\overline{y}_n}(x) \text{ for a.c. } x \in X \tag{11.2}$$

and by the definition of u we obtain that

$$u_g(x) = u(x, g(x)) \text{ for a.e. } x \in X.$$

We have now proved that the function u has the property that

$$f(g, A) = \int_A u(x, g(x))d\mu \text{ for every } g \in \mathcal{G} \text{ and } A \in \mathcal{A}.$$

In order to prove that u is weakly quasi-continuous, we consider a sequence (g_n) in $\mathcal{G}$ and $g \in \mathcal{G}$ such that $g_n \to g$ a.e. We have prove that $u(x, g_n(x)) \to u(x, g(x))$ a.e. $x \in X$. However since for all $h \in \mathcal{G}, u(x, h(x)) = u_h(x)$ a.e. $x \in X$, this amounts to prove that $u_{g_n}(x) \to u_g(x)$ a.e. But this can be proved in exactly the same way as the proof of lemma 23 replacing throughout the functions $(\overline{y}_n)$ by the functions (g_n) ■

It is now easy to obtain integral representations also for the representation which were themselves corollaries of theorem 26 page 95.

1. Let $(X, \mathcal{A}, \mu)$ be a positive σ-finite measure space, and (Y, d) a separable metric space. $\mathcal{G}$ is a set of measurable functions defined on X with values in Y, and Q a subset of $\mathcal{G}$.

Let $\mathcal{U}$ be the finest uniformity on Y compatible with the topology on Y and let t be the topology of $\mathcal{U}$-uniform convergence on $\mathcal{G}$.

Corollary 18 *Let $(\mathcal{G}, Q)$ be an independent and connected mixture with respect to the σ-algebra $\mathcal{A}$ containing more than two non-null sets and containing the constant functions. Assume that $(\mathcal{G}, Q)$ is continuous with respect to the measure μ and that the order topology is coarser than t. Then there exists a weakly quasi-continuous* *function* $u : X \times Y \to \mathbb{R}$ *such that, when the function* $f : \mathcal{G} \times \mathcal{A} \to \mathbb{R}$ *is defined by*

$$f(g, A) = \int_A u(x, g(x)) d\mu,$$

then

$$f(g, X) > 0 \Longleftrightarrow g \in Q$$

Proof. See theorem 52 and remark 37 ■

2. Let $(X, \mathcal{A}, \mu)$ be a positive σ-finite measure space, and $(Y \times Z, d)$ a separable metric space. $\mathcal{G} \times \mathcal{H}$ is a set of measurable functions defined on X with values in $Y \times Z$, $\mathcal{P}$ a relation on $\mathcal{G} \times \mathcal{H}$ and $graph\mathcal{P} = Q$ is a subset of $\mathcal{G} \times \mathcal{H}$.

Let $\mathcal{U}$ be the finest uniformity on $Y \times Z$ compatible with the topology on $Y \times Z$ and let t be the topology of $\mathcal{U}$-uniform convergence on $\mathcal{G} \times \mathcal{H}$.

Corollary 19 *Let* $(\mathcal{G} \times \mathcal{H}, \mathcal{P})$ *be an independent and connected mixture with respect to the* σ*-algebra* $\mathcal{A}$ *containing more than two non-null sets and containing the constant functions.* Q *determines a preorder on* $\mathcal{G} \times \mathcal{H}$*. Assume that* $(\mathcal{G} \times \mathcal{H}, \succsim)$ *is continuous with respect to the measure* μ *and that the order topology determined by* Q *via* $\succsim$ *is coarser than* t*.Then there exists a weakly quasi-continuous function* $u : X \times Y \times Z \to \mathbb{R}$ *such that, when the function* $f : \mathcal{G} \times \mathcal{A} \to \mathbb{R}$ *is defined by*

$$f(g, h, A) = \int_A u(x, g(x), h(x)) d\mu,$$

then

$$F(g, h) = f(g, h, X) = \int_X u(x, g(x), h(x)) d\mu > 0 \Longleftrightarrow g \in \mathcal{P}(h)$$

Proof. Special case of corollary 18 for Y replaced by $Y \times Z$ and $\mathcal{G}$ by $\mathcal{G} \times \mathcal{H}$ ■

3. Let $(X, \mathcal{A}, \mu)$ be a positive σ-finite measure space, and (Y, d) a separable metric space. $\mathcal{G}$ is a set of measurable functions defined on X with values in Y, and $\mathcal{P}$ a elation on $\mathcal{G}$.

Let $\mathcal{U}$ be the finest uniformity on Y compatible with the topology on Y and let t be the topology of $\mathcal{U}$-uniform convergence on $\mathcal{G}$.

Corollary 20 *Let $(\mathcal{G}, \mathcal{P})$ be an independent and connected mixture with respect to the σ-algebra $\mathcal{A}$ containing more than two non-null sets and containing the constant functions. graph$\mathcal{P}$ determines a preorder and an order topology on $\mathcal{G} \times \mathcal{G}$. Assume that $(\mathcal{G} \times \mathcal{G}, \succsim)$ is continuous with respect to the measure μ and that the order topology is coarser than t. Then there exists a weakly quasi-continuous function $u : X \times Y \to \mathbb{R}$ such that, when the function $f : \mathcal{G} \times \mathcal{A} \to \mathbb{R}$ is defined by*

$$f(g, h, A) = \int_A u(x, g(x), h(x)) d\mu,$$

then

$$f(g, h, X) > 0 \Longleftrightarrow g \in \mathcal{P}(h)$$

Proof. Put $Z = Y$ and $\mathcal{H} = \mathcal{G}$ in corollary 19 ■

Remark 50 *The assumption that the order topology is coarser than t can of course be replaced by the stronger assumption that the order topology is coarser than uniform convergence on $\mathcal{G}$ corresponding to the metric d. However this is an unpleasant assumption since equivalent metrics on Y may induce different topologies on $\mathcal{G}$. Also, the assumption that the order topology is coarser than t, can of course be replaced by the stronger assumption that for every sequence (g_n) in $\mathcal{G}$ and $g \in \mathcal{G}$ with $g_n \to g$ a.e., we have that $g_n \to g$ in the order topology.*

Remark 51 *The following theorems (53 and 55) have obvious corollaries corresponding to the corollaries above.*

11.3 Continuity and boundedness of u

In theorem 52 we obtained a measurable weakly quasi-continuous function u, such that the function $f(\cdot, A)$ defined by

$$f(g, A) = \int_A u(x, g(x)) d\mu$$

is strictly monotonic. In this section we shall derive further properties of the function u.

Theorem 53 *Let $(\mathcal{G}, \precsim)$ be as in theorem 52 and assume moreover that $\mathcal{G}$ is a σ-mixture.*

Then there exists a function $u : X \times Y \to \mathbb{R}$ having the properties from theorem 52, but moreover

(i) u is strongly quasi-continuous and

(ii) the function

$$S : X \to \mathbb{R} \cup \{+\infty\}$$

defined by

$$S(x) = \sup_{y \in Y} |u(x,y)|$$

is infinite only on a finite number of atoms, and is integrable on the remaining part of X.

Proof. Let $u, \widetilde{u} : X \times Y \to \mathbb{R}$ be the functions defined in the proof of theorem 52 i.e. $\widetilde{u}(x,y) = \limsup_{y_n \to y} u_{\overline{y}_n}(x)$ and

$$u(x,y) = \begin{cases} \widetilde{u}(x,y) & \text{if } \widetilde{u}(x,y) < +\infty \\ 0 & \text{if } \widetilde{u}(x,y) = +\infty \end{cases}$$

Let moreover $r : X \to Y$ be defined by $r(x) = \sup_{y \in Y} \widetilde{u}(x,y)$. We first prove that r has the properties stated in (ii).

First we remark that for a.e. $x \in X, r(x) = \sup_n u_{\overline{y}_n}(x)$.

Indeed

$$\sup_n \widetilde{u}(x, y_n) \leq r(x) \leq \sup_n u_{\overline{y}_n}(x) \text{ for all } x \in X$$

and from lemma 23 we have that for all n,

$$\widetilde{u}(x, y_n) = u_{\overline{y}_n}(x) \text{ a.e. } x \in X.$$

Therefore

$$\sup_n \widetilde{u}(x, y_n) = \sup_n u_{\overline{y}_n}(x) a.e. \tag{11.3}$$

and consequently

$$r(x) = \sup_n u_{\overline{y}_n}(x) a.e. \tag{11.4}$$

Especially this implies that r is measurable.

Now consider the set

$$S = \{x \in X | r(x) = +\infty\}$$

If S contains an atomless part T with $0 < \mu(T) < +\infty$, or countably many atoms, then we can construct a function $g \in \mathcal{G}$ such that $f(g, X) = +\infty$, which contradicts $f : \mathcal{G} \times \mathcal{A} \to \mathbb{R}$.

First assume S contains an atomless part T with $0 < \mu(T) < +\infty$, and let (T_i) be a partition of T with $\mu(T_i) = \frac{1}{2^i}\mu(T)$. Define $t : X \to \mathbb{R}$ by

$$t(x) = \begin{cases} 1 & \text{if} \quad x \notin T \\ \inf\{k | u_{\overline{y}_k}(x) \geq 2^i\} & \text{if} \quad x \in T_i \end{cases}$$

and consider the function $g : X \to Y$ given by $g(x) = y_{t(x)}$ for all $x \in X$. Since $\mathcal{G}$ is a σ-mixture containing the constant functions, $g \in \mathcal{G}$, and by the construction of g we have that

$$f(g, T) \geq \sum_i 2^i \frac{1}{2^i}\mu(T) = +\infty$$

If S contains countably many atoms $(T_i)_{i=1}^{\infty}$, then we in the same way get a contradiction by considering the function g, with $g(x) = y_{t(x)}$, where $t(x) = \inf\{k | u_{\overline{y}_k}(x) \geq \frac{1}{\mu(T_i)}\}$.

This proves that S, except for a null set in $(X, \mathcal{A}, \mu)$, is the union of a finite number of atoms.

We moreover have to prove that r is integrable outside S. Let $h : X \to \mathbb{R}$ be any strictly positive integrable function and define $t : X \to \mathbb{R}$ by

$$t(x) = \begin{cases} 1 & \text{if} \quad x \in S \\ \inf\{k | u_{\overline{y}_k}(x) > r(x) - h(x)\} & \text{if} \quad x \notin S \end{cases}$$

and correspondingly $g : X \to Y$ by $g(x) = y_{t(x)}$ for all $x \in X$. Then $g \in \mathcal{G}$ and

$$\int_{S^c} r d\mu \leq \int_{S^c} u_g d\mu + \int_{S^c} h d\mu < +\infty$$

which shows that r is integrable on the complement of S.

We have now proved the properties in (ii) for the function r.

Clearly also the function $\widetilde{r}$, defined by $\widetilde{r}(x) = \sup_{y \in Y} -\widetilde{u}(x,y)$, shall have these properties and thereby the function $\widetilde{S}$, where $\widetilde{S}(x) = \sup_{y \in Y} |\widetilde{u}(x,y)|$. However this implies by the definition of u that (ii) is true.

There remains to prove (i). Consider an arbitrary sequence (g_n) in $\mathcal{G}$ and $g \in \mathcal{G}$ with $g_n \to g$ a.e. We know from theorem 52 that $u(\cdot, g_n)) \to u(\cdot, g(\cdot))$ a.e. But from (ii) we have that the sequence $(u(\cdot, g_n))$ is bounded by the integrable function S except on the union of finitely many atoms, and consequently that $(u(\cdot, g_n(\cdot))$ converges to $u(\cdot, g(\cdot))$ in $L_1(X, \mathcal{A}, \mu)$ ∎

Theorem 54 *Let Y be a complete separable metric space, $(X, \mathcal{A}, \mu)$ a complete positive σ-finite measure space, and $\mathcal{G}$ the set of all measurable functions from X to Y. Moreover assume that $(\mathcal{G}, \succsim)$ has the properties specified in theorem 52.*

Then there exists a function $u : X \times Y \to \mathbb{R}$ having the properties specified in theorem 52 and moreover, for almost all $x \in X$ the function $u(x, \cdot) : Y \to \mathbb{R}$ is continuous.

Proof. First prove that except for a null set in $(X, \mathcal{A}, \mu)$,

$$\mathop{limsup}_{y_n \to y} u_{\overline{y}_n}(x) = \mathop{liminf}_{y_n \to y} u_{\overline{y}_n}(x) \text{ for all } y \in Y.$$

Let $\widetilde{u}, u$ be the functions defined in the proof of theorem 52 and define the function $v : X \times Y \to \mathbb{R}$ by $v(x,y) = \mathop{liminf}_{y_n \to y} u_{\overline{y}_n}(x)$.

Moreover let

$$H = \{(x,y) \in X \times Y | \widetilde{u}(x,y) > v(x,y)\}$$

Since $\widetilde{u}$ and v are measurable functions, H is measurable. But then, since Y is a complete, separable metric space and $(X, \mathcal{A}, \mu)$ is complete, $N = \text{proj}_X H \in \mathcal{A}$ and there exists a measurable function $g : N \to Y$ such that $\{(x, g(x)) | x \in N\} \subset H$. Let us now define the function $h : X \to Y$ by

$$h(x) = \begin{cases} y_1 & \text{if } x \notin N \\ g(x) & \text{if } x \in N \end{cases}$$

Since $\mathcal{G}$ is the set of all measurable functions, then $h \in \mathcal{G}$. Using lemma 23 we therefore have

$$u_h(x) = \lim_{y_n \to h(x)} u_{\overline{y}_n}(x) \text{ for a.e.} x \in X$$

and therefore that

$$\widetilde{u}(x, h(x)) = v(x, h(x)) \text{ for a.e.} x \in X.$$

But then $\mu(N) = 0$, i.e. for almost all $x \in X$,

$$\mathit{limsup}_{y_n \to y} u_{\overline{y}_n}(x) = \mathit{liminf}_{y_n \to y} u_{\overline{y}_n}(x) \text{ for all } y \in Y,$$

which was to be proved.

In order to prove that $u(x, \cdot) : Y \to \mathbb{R}$ is continuous for almost all $x \in X$, we first notice that the definition of H imply that $\widetilde{u}(x, y) = u(x, y)$ for all $y \in Y$, if $x \notin N$. Moreover letting

$$N_n = \{x \in X | u_{\overline{y}_n}(x) \neq u(x, y_n)\}$$

we have from theorem 52 that $\mu(N_n) = 0$ for all n. Now define

$$N' = \bigcup_{n=1}^{\infty} N_n \cup N$$

and notice that $\mu(N') = 0$.

We shall show that $u(x, \cdot) : Y \to \mathbb{R}$ is continuous for all $x \notin N'$. Let $x \in X \backslash N'$ and $y_0 \in Y$, and consider $\varepsilon > 0$. Since

$$u(x, y_0) = \lim_{y_n \to y_0} u_{\overline{y}_n}(x) = \lim_{y_n \to y_0} u(x, y_n)$$

we can choose δ, such that

$$d(y_n, y_0) < \delta \Longrightarrow |u(x, y_0) - u(x, y_n)| < \frac{\varepsilon}{2}$$

Now let $y' \in Y$ be such that $d(y', y_0) < \delta$. Since

$$\lim_{y_k \to y'} u(x, y_k) = u(x, y')$$

we get that

$$|u(x, y_0 - u(x, y')| \leq \frac{\varepsilon}{2} < \varepsilon$$

This proves that $u(x, \cdot)$ is continuous ■

11.4 Existence of $u : X \times Y \to \mathbb{R}$ when $\mathcal{G}$ is a set of measurable selections

In sections 11.2 and 11.3 we assumed that $\mathcal{G}$ contained all constant functions from X to Y. In this section we shall assume that the set $\mathcal{G}$ is given as all measurable selections from a measurable correspondence.

Let

$$\varphi : X \to 2^Y$$

be a measurable correspondence i.e.

$$graph\varphi = \{(x, y) \in X \times Y | y \in \varphi(x)\}$$

is measurable. We shall let $\mathcal{G}$ be the set of measurable selection from φ i.e.

$$\mathcal{G} = \{g : X \to Y | g \text{ measurable and } g(x) \in \varphi(x) \ a.e.\ x \in X\}.$$

Clearly $\mathcal{G}$ is a σ-mixture.

Remark 52 *The set $\mathcal{G}$ considered in section 11.3, namely all measurable functions from X to Y, is identical to the set of measurable selections from the measurable correspondence $\varphi : X \to 2^Y$ defined by $\varphi(x) = Y$ for all $x \in X$.*

Theorem 55 *Let Y be a complete, separable metric space and let $(X, \mathcal{A}, \mu)$ be a complete positive σ-finite measure space. Moreover let $\mathcal{G}$ be the set of all measurable selections from a measurable correspondence φ from X to Y.*

Assume that $(\mathcal{G}, \succsim)$ is an independent and connected mixture with respect to the σ-algebra $\mathcal{A}$ and that $\mathcal{A}$ has more than two non-null sets. Moreover assume that $(\mathcal{G}, \succsim)$ is continuous with respect to the measure μ and that the order topology is coarser than t.

Then there exists a measurable function

$$u : graph\varphi \to \mathbb{R}$$

such that for almost all $x \in X$

$$u(x, \cdot) : \varphi(x) \to \mathbb{R}$$

is continuous, and such that $f(\cdot, A)$ *is strictly monotonic on* $\mathcal{G}|A$, *when*

$$f : \mathcal{G} \times \mathcal{A} \to \mathbb{R}$$

is defined by

$$f(g, A) = \int_A u(x, g(x))d\mu.$$

Proof. As φ has a measurable graph, Y is a complete separable metric space, and $(X, \mathcal{A}, \mu)$ is complete, there exists a sequence (g_n) of measurable functions from X to Y such that $g_n \in \mathcal{G}$ for all n and for all $x \in X$, $\{g_n(x) \,|n \in N\}$ is dense in $\varphi(x)$; (see Aumann (1969) [12]).

The proof of the theorem is now analogous to the proofs of theorem 52 and theorem 54 replacing the constant functions $(\overline{y}_n)_{n=1}^{\infty}$by the functions $(g_n)_{n=1}^{\infty}$ ∎

11.5 Notes

1. This chapter is based on B. Grodal and J.-F. Mertens: "Integral representation of utility functions." C.O.R.E. discussion paper no. 6823 [91].

2. In order to get the existence of measurable function $u : X \times Y \to \mathbb{R}$ such that the function $f : \mathcal{G} \times \mathcal{A} \to \mathbb{R}$ defined by $f(g, A)) = \int_A u(x, g(x))d\mu$ is strictly monotonic on $\mathcal{G}|A$ some continuity condition on the preordering $\succsim$ is needed. Consider the following total preordering on the set of all measurable functions from a finite positive measure space $(X, \mathcal{A}, \mu)$ to the separable metric space Y :

   $$g_1 \succsim g_2 \Leftrightarrow \sum_{y \in Y} \mu\{x \in X | g_1(x) = y\} \geq \sum_{y \in Y} \mu\{x \in X | g_2(x) = y\}$$

   The sum is well defined since at most countably many of the sets have positive measure.

   It is easily seen that this preordering can not be represented by means of a function $u : X \times Y \to \mathbb{R}$. Indeed in order that f

is strictly monotonic on the restriction to the step functions in $\mathcal{G}$ then, except for a nullset in X, $u(\cdot,\cdot)$ has to be a constant function, but then f is not strictly monotonic on $\mathcal{G}|A$.

3. The theorems in this chapter have two classical interpretations in economic theory. Firstly, $X = [0,\infty[$ can be interpreted as time and Y as the consumption space. Consequently, $\mathcal{G}$ is interpreted as a set of consumption programs over time. (See chapters 12 and 15). Secondly, one can interpret $(X, \mathcal{A}, \mu)$ as a probability space of events, and Y as the space of consequences. In this case $\mathcal{G}$ is a set of programs over states of nature, actions, where for each action $g \in \mathcal{G}$ and each state of nature $x \in X$ is specified a consequence $g(x) \in Y$.

   In both cases theorem 52 states conditions on $(\mathcal{G}, \succsim)$ which allow us to define for each $x \in X$ i.e. either for each time or for each state of nature, a utility function $u(x,\cdot) : Y \to \mathbb{R}$ such that we obtain a utility function f for $(\mathcal{G}, \succsim)$ by letting the utility of a program $g \in \mathcal{G}$ be the integral with respect to μ of the utilities obtained for each $x \in X$. In the case where $(X, \mathcal{A}, \mu)$ is interpreted as a probability space of events this especially means that the utility of $g \in \mathcal{G}$ is the expected utility.

4. An interpretation of the correspondence φ in section 11.4, corresponding to the time interpretation of $(X, \mathcal{A}, \mu)$ is that $\varphi(x)$ is the consumption set at time x.

5. The problem treated in this chapter is related to work by Savage (1954) [160], Fishburn (1967b) [74], and Villegas (1964) [175]. E.g. Savage consider a totally preordered set $(\mathcal{G}, \succsim)$, where the functions $g \in \mathcal{G}$ is defined on a σ-algebra $(X, \mathcal{A})$. He gives conditions which imply the existence of a probability measure P on $(X, \mathcal{A})$ and a function $\overline{u} : Y \to \mathbb{R}$ such that

$$f(g) = \int \overline{u}(g(x))dP \tag{11.5}$$

   is strictly monotonic on $(\mathcal{G}, \succsim)$. This problem differs from the problem treated here in two respects. Firstly, we start with a measure space $(X, \mathcal{A}, \mu)$ and want μ to be the measure with respect to which we integrate. Secondly we allow different utility functions $u(x, \cdot)$ for $x \in X$.

Let us assume that the measure P obtained in [160] is continuous with respect to the given measure μ. (It is easy to state conditions on $(\mathcal{G}, \succsim)$ which imply this). Then the utility function (11.5) can be rewritten $f(g) = \int \overline{u}(g(x))\alpha(x)d\mu$, where $\alpha : X \to \mathbb{R}$ is measurable. Compared with theorem 52 this means that $u : X \times Y \to \mathbb{R}$ has the special form $u(x, y) = \overline{u}(y)\alpha(x)$. For this problem see also chapter 12.

12

Special integral representations by Birgit Grodal

12.1 Introduction

In chapter 11 we gave conditions on $(\mathcal{G}, \succsim)$ which implied the existence of a measurable function

$$u : X \times Y \to \mathbb{R}$$
$$\text{such that}$$
$$U(g) = \int_X u(x, g(x)) d\mu$$

is an order-homomorphism on $(\mathcal{G}, \succsim)$. Further conditions implied continuity properties of u. In this chapter we give still more special conditions on $(\mathcal{G}, \succsim)$ which allow us to obtain functions u with special properties.

First we give conditions on $(\mathcal{G}, \succsim)$ which imply that there exists a continuous function $\overline{u} : Y \to \mathbb{R}$ and a measurable function $\beta : X \times \overline{u}(Y) \to \mathbb{R}$ such that

$$\beta(x, \cdot) : \overline{u}(Y) \to \mathbb{R}$$

is strictly monotonic and such that

$$U(g) = \int_X \beta(x, \overline{u}(g(x))) d\mu$$

is an order-homomorphism on $(\mathcal{G}, \succsim)$. In this case there exists a fixed preordering on Y, namely the one represented by the function $\overline{u}$, but in order to get the order-homomorphism on $(\mathcal{G}, \succsim)$ we need for a given $x \in X$ to use the representation $\beta(x, \overline{u}(\cdot))$ of that preordering on Y.

Secondly we find conditions such that the function

$$\beta : X \times \overline{u}(Y) \to \mathbb{R}$$
has the form
$$\beta(x, \overline{u}(y)) = \overline{u}(y)\alpha(x)$$
where $\alpha : X \to \mathbb{R}$

So

$$U(g) = \int_X \overline{u}(y)\alpha(x) d\mu$$

Thirdly in the case where $X = T = [0, \infty[$ we give conditions for the existence of $\delta > 0$ such that $\beta(t, \overline{u}(y)) = \overline{u}(y)e^{-\delta t}$, i.e. the function $\alpha : T \to \mathbb{R}$ has the special form $e^{-\delta t}$. So

$$U(g) = \int_X \overline{u}(y)e^{-\delta t} d\mu$$

12.2 $f(g, A) = \int_A \beta(x, \overline{u}(g(x)))d\mu$

Let $(X, \mathcal{A}, \mu)$ be a positive σ-finite measure space and let (Y, d) be a separable metric space. As in chapter 11, $\mathcal{G}$ is a set of measurable functions defined on X with values in Y, and $\succsim$ is a total preordering on $\mathcal{G}$. Also as in chapter 11, let $\mathcal{U}$ be the finest uniformity on Y compatible with the topology on Y and let t be the topology of $\mathcal{U}$-uniform convergence on $\mathcal{G}$.

We shall assume that $(\mathcal{G}, \succsim)$ is an independent mixture containing the constant functions. As $(\mathcal{G}, \succsim)$ is an independent mixture, there is for each $A \in \mathcal{A}$ a uniquely determined total preordering $\succsim_A$ on $\mathcal{G}|A$:

$$\forall g, g' \in \mathcal{G} : g|A \succsim_A g'|A \Leftrightarrow (g|A, h|A^c) \succsim (g'|A, h|A^c)$$

where h is an arbitrary function in $\mathcal{G}$.

Since the constant functions moreover are assumed to be contained in $\mathcal{G}$ we can, for each $A \in \mathcal{A}$, use $\succsim_A$ to define a total preordering on Y, a preordering we again denote $\succsim_A$. Namely, let $\succsim_A$ on Y be defined by:

$$\forall y, y' \in Y : y \succsim_A y' \Leftrightarrow \overline{y}|A \succsim_A \overline{y}'|A.$$

Definition 140 (($\mathcal{G}, \succsim$) independent with respect to $\mathcal{A}$) *Let $(\mathcal{G}, \succsim)$ be an independent mixture containing the constant functions and let $\succsim_A$ be the induced preordering on Y by $A \in \mathcal{A}$. We shall say that $(Y, \succsim_A)$ is independent with respect to $\mathcal{A}$ if for every non-null sets $A, B \in \mathcal{A}$*

$$\succsim_A = \succsim_B$$

Theorem 56 *Let $(X, \mathcal{A}, \mu)$ a complete positive σ-finite measure space, Y be a complete, connected, and separable metric space, and $\mathcal{G}$ the set of all measurable functions from X to Y. Assume that $(\mathcal{G}, \succsim)$ has the properties, specified in theorem 52 and moreover that $(Y, \succsim_A)$ is independent with respect to $\mathcal{A}$. Then there exists a continuous function $\overline{u} : Y \to \mathbb{R}$ and a measurable function*

$$\beta : X \times \overline{u}(Y) \to \mathbb{R}$$

such that, when the function

$$f : \mathcal{G} \times \mathcal{A} \to \mathbb{R}$$

is defined by

$$f(g, A) = \int_A \beta(x, \overline{u}(f(x)))d\mu,$$

then $f(\cdot, A)$ is strictly monotonic on $\mathcal{G}|A$ for all $A \in \mathcal{A}$. Moreover except for x belonging to a null set with respect to $(\mathcal{G}, \succsim)$,

$$\beta(x, \cdot) : \overline{u}(Y) \to \mathbb{R}$$

is a strictly monotonic function.

Proof. Let $(y_n)_{n=1}^{\infty}$ be a dense subset in Y, and let $f : \mathcal{G} \times \mathcal{A} \to \mathbb{R}$ be a function determined by theorem 26. Using the remark 31 page 95 we can assume that $f(\overline{y}_1, A) = 0$ for all $A \in \mathcal{A}$.

Now we use the construction from the proof of theorem 52 to define a measurable function $u : X \times Y \to \mathbb{R}$ such that for every $g \in \mathcal{G}$ and $A \in \mathcal{A}$,

$$f(g, A) = \int_A u(x, g(x))d\mu.$$

Since all assumptions of theorem 55 are satisfied there exists $N' \in \mathcal{A}, \mu(N') = 0$, such that for all $x \in X \backslash N'$,

$$u(x, \cdot) : Y \to \mathbb{R} \text{ is continuous.}$$

We shall show that the function $u : X \times Y \to \mathbb{R}$ has the form $u(x,y) = \beta(x, \overline{u}(y))$, where $\overline{u}$ and β are as stated in the theorem.

First we remark that there exists a null set E in $(\mathcal{G}, \succsim)$, such that if E' is a null set in $(\mathcal{G}, \succsim)$ then $\mu(E' \backslash E) = 0$. Indeed this follows since $(X, \mathcal{A}, \mu)$ is σ-finite. Moreover we remark that since for all $n \in N, u(\cdot, y_n) : X \to \mathbb{R}$ is a Radon Nikodým derivative of the measure $f(\overline{y}_n, \cdot) : \mathcal{A} \to \mathbb{R}$, and $(Y, \succsim_A)$ is independent with respect to $\mathcal{A}$, then there exists $S \in \mathcal{A}, \mu(S) = 0$, such that for every $n, m \in N$.

$$\begin{gathered} f(y_n, X) < f(y_m, X) \Rightarrow \forall x \in X \backslash S \cup E : u(x, y_n) < u(x, y_m) \\ \text{and} \\ f(y_n, X) = f(y_m, X) \Rightarrow \forall x \in X \backslash S \cup E : u(x, y_n) = u(x, y_m) \end{gathered} \tag{12.1}$$

Let $E' = E \cup S \cup N'$. Then $E' \in \mathcal{A}$ and E' is a null set in $(\mathcal{G}, \succsim)$. Clearly we can assume that $u(x,y) = 0$ for every $x \in E'$ and $y \in Y$.

We shall now prove that for every $y, y' \in Y$

$$f(\overline{y}, X) < f(\overline{y}', X) \Rightarrow \forall x \in X \backslash E' : u(x,y) < u(x,y') \tag{12.2}$$

First we remark that the definition of

$$u(\cdot, \cdot) : X \times Y \to \mathbb{R}$$

implies that we can assume

$$u(x,y) = \limsup_{y_n \to y} u(x, y_n) \text{ for all } y \in Y \text{ and } x \in X$$

(see proof of theorem 52). Since $f(\cdot, X) : Y \to \mathbb{R}$ is continuous and Y is connected, (12.1) consequently implies that for all $y, y' \in Y$

$$f(\overline{y}, X) < f(\overline{y}', X) \Rightarrow \forall x \in X \backslash E' : u(x,y) \leq u(x,y') \tag{12.3}$$

However still using that $\{f(\overline{y}, X) | y \in Y\}$ is connected, (12.3) and (12.1) imply (12.2). Now assume that there exists $y, y' \in Y$ such that $f(\overline{y}, X) = f(\overline{y}', X)$ and $u(\overline{x}, y) \neq u(\overline{x}, y')$ for some $\overline{x} \in X \backslash E'$. By symmetry we can assume $u(\overline{x}, y) < u(\overline{x}, y')$. Since Y is connected and $u(\overline{x}, \cdot) : Y \to \mathbb{R}$ is continuous, it is possible to find $n, m \in N$ such that

$$u(\overline{x}, y) < u(\overline{x}, y_n) < u(\overline{x}, y_m) < u(\overline{x}, y')$$

However by (12.1) and (12.3) this implies

$$f(\overline{y}, X) \leq f(\overline{y}_n, X) < f(\overline{y}_m, X) \leq f(\overline{y}', X)$$

and consequently gives a contradiction.

We have now proved that for all $y, y' \in Y$

$$\begin{gathered} f(\overline{y}, X) < f(\overline{y}', X) \Leftrightarrow \forall x \in X \backslash E' : u(x, y) < u(x, y') \\ \text{and} \\ f(\overline{y}, X) = f(\overline{y}', X) \Leftrightarrow \forall x \in X \backslash E' : u(x, y) = u(x, y') \end{gathered} \qquad (12.4)$$

We now define the function

$$\overline{u} : Y \to \mathbb{R} \text{ by } \overline{u}(y) = f(\overline{y}, X)$$

Then $\overline{u}$ is continuous.

Moreover since $f(\overline{y}, X) = f(\overline{y}', X)$ implies that $u(x, y) = u(x, y')$ for all $x \in X$, we can define the function $\beta : X \times \overline{u}(Y) \to \mathbb{R}$ by $\beta(x, r) = u(x, y)$ where y is chosen such that $\overline{u}(y) = r$. We have to show that β is measurable.

Let $S = \sup_{y \in Y} f(\overline{y}, X)$. If there exists $y \in Y$ such that $f(\overline{y}, X) = S$, let $y' \in Y$ be a fixed element in Y having this property.

Now we have for any $\lambda \in \mathbb{R}$ that

$$\{(x, r) \in X \times \mathbb{R} | \beta(x, r) < \lambda\} =$$

$$\bigcup_{n=1}^{\infty} (\{x \in X | u(x, y_n) < \lambda\} \times \{r \in \mathbb{R} | r < f(\overline{y}_n, X)\}) \cup T$$

where

$$T = \begin{cases} \emptyset & \text{if for all } y \in Y : S > f(\overline{y}, X) \\ \{x \in X | u(x, y') < \lambda\} \times \{S\} & \text{if } y' \text{ exists} \end{cases}$$

Consequently β is measurable.

To complete the proof we just have to remark that (12.4) implies that for every $x \in X \backslash E'$ we have that $\beta(x, \cdot)$ is a strictly monotonic function ■

12.3 $f(g, A) = \int_A \overline{u}(g(x))\alpha(x)d\mu$

We shall now make further assumptions on $(\mathcal{G}, \succsim)$ in order to get more properties of the function β obtained above.

Let us still assume that $(\mathcal{G}, \succsim)$ is an independent and connected mixture with respect to the σ-algebra $\mathcal{A}$ which contains more than two non-null sets and that $\mathcal{G}$ contains the constant functions. Also we assume that Y is connected and that the order topology is coarser than t. From theorem 24 we know it is possible for any $A \in \mathcal{A}$ to define a mean operation $\circ_A$ on the set of equivalence classes in $(\mathcal{G}|A, \succsim_A)$, such that $(\mathcal{G}|A/\sim_A, \succsim_A, \circ_A)$ is a commutative mean groupoid.

Now let $A \in \mathcal{A}$. We saw above that we could introduce a total preordering $\succsim_A$ on Y by using the constant functions in $\mathcal{G}|A$. We now remark that since Y is connected, the mean operation $\circ_A$ on $\mathcal{G}|A$ also allows us to define a mean operation on the equivalence classes for $\succsim_A$ on Y. Indeed let $y, y' \in Y$ and consider the mean $\overline{y}|A \circ_A \overline{y}'|A$ in $\mathcal{G}|A/\sim_A$. Since Y is connected then there exists $y'' \in Y$ such that $\overline{y}''|A \in \overline{y}|A \circ_A \overline{y}'|A$.

Now we define $\circ_A$ on Y by letting $y \circ_A y'$ be the equivalence class in $Y/\sim_A$ containing y''. Clearly $(Y/\sim_A, \succsim_A, \circ_A)$ is a commutative mean groupoid.

Also we remark that if we assume that $(Y, \succsim_A)$ is independent with respect to $\mathcal{A}$, then

$$(Y\backslash \sim_A, \succsim_A) = (Y\backslash \sim_B, \succsim_B)$$

for all non-null sets $A, B \in \mathcal{A}$. We now define

Definition 141 (**$(Y/\sim_A, \succsim_A, \circ_A)$ independent w. r. t. $\mathcal{A}$**)

$$(Y/\sim_A, \succsim_A, \circ_A)$$

is independent with respect to $\mathcal{A}$ if $(Y, \succsim_A)$ is independent with respect to $\mathcal{A}$ and if for every non-null $A, B \in \mathcal{A}$

$$\circ_A = \circ_B$$

We are now able to state and prove the following

Theorem 57 *Let Y be a complete, connected, and separable me tric space, $(X, \mathcal{A}, \mu)$ a complete positive σ-finite measure space.*

Moreover let $(\mathcal{G}, \succsim)$ be an independent and connected mixture with more than two non-null sets and containing the constant functions. Assume that $(\mathcal{G}, \succsim)$ is continuous with respect to μ that the order topology is coarser than t, and that $(Y/\sim_A, \succsim_A, \circ_A)$ is independent with respect to $\mathcal{A}$. Then there exists a continuous function $\overline{u}: Y \to \mathbb{R}$ and a measurable function $\alpha : X \to \mathbb{R}$ such that, when the function $f : \mathcal{G} \times \mathcal{A} \to \mathbb{R}$ is defined by

$$f(g,A) = \int_A \overline{u}(g(x))\alpha(x)d\mu$$

then $f(\cdot, A)$ is strictly monotonic on $\mathcal{G}|A$ for all $A \in \mathcal{A}$.

Proof. Let $(y_n)_{n=1}^{\infty}$ be a dense subset in Y and let $f : \mathcal{G} \times \mathcal{A} \to \mathbb{R}$ be determined by theorem 26.

As in the proof of theorem 56 we can assume that $f(\overline{y}_1, A) = 0$ for all $A \in \mathcal{A}$.

Let A be an arbitrary non-null set in $\mathcal{A}$.

First we remark that since $(Y, \succsim_A)$ is independent with respect to $\mathcal{A}$ then, if $y, y' \in Y$ and $f(\overline{y}, X) = f(\overline{y}', X)$, then $f(\overline{y}, E) = f(\overline{y}', E)$ for all $E \in \mathcal{A}$. Now let $I = \{f(\overline{y}, X) | y \in Y\}$ and define the mapping

$$\gamma_A : I \to \mathbb{R} \text{ by } \gamma_A(r) = f(\overline{y}, A)$$

where y has been chosen such that $f(\overline{y}, X) = r$. We shall prove that there exists $\alpha_A \in \mathbb{R}$ such that

$$\gamma_A(r) = \alpha_A r \quad \text{for every } r \in I \tag{12.5}$$

To do this we show that γ_A is continuous and satisfies the following equation

$$\gamma_A\left(\frac{r_1 + r_2}{2}\right) = \frac{\gamma_A(r_1) + \gamma_A(r_2)}{2} \quad \text{for every } \ r_1, r_2 \in I \tag{12.6}$$

Indeed γ_A is continuous, since $(Y, \succsim_A)$ is assumed to be independent with respect $\mathcal{A}$ and since for any $E \in \mathcal{A}$ the restriction of $f(\cdot, E)$ to $(Y/\sim_E)$ is an order isomorphism. To see that γ_A satisfies equation (12.6) we first remark that for any non-null set $E \in \mathcal{A}$, the definition of f implies that for any $y, y', y" \in Y$

$$f(\overline{y}, E) = \frac{f(\overline{y}', E) + f(\overline{y}", E)}{2} \text{ if and only if } y \in y' \circ_E y"$$

This follows from the definition of $\circ_E$ and f, see chapter 6.

Now let $r_1, r_2 \in I$ and choose $y', y" \in Y$ such that $f(\overline{y}', X) = r_1$ and $f(\overline{y}", X) = r_2$. Moreover define $y \in Y$ such that $y \in y' \circ_X y"$. As remarked above we have that

$$f(\overline{y}, X) = \frac{f(\overline{y}', X) + f(\overline{y}", X)}{2}$$

Consequently we get that

$$\gamma_A\left(\frac{r_1 + r_2}{2}\right) = f(\overline{y}, A)$$

Moreover since $(Y/\sim_A, \succeq_A, \circ_A)$ is independent with respect to $\mathcal{A}$, we have that $y \in y' \circ_A y"$and therefore again using the remark above that

$$f(\overline{y}, A) = \frac{f(\overline{y}', A) + f(\overline{y}", A)}{2}$$

Finally this implies (12.6).

Now it is well known (see e.g. Aczél (1966) [5], p. 43) that (12.6) implies that γ_A is of the form

$$\gamma_A(r) = \alpha_A r + a_A \text{ where } \alpha_A, a_A \in \mathbb{R}.$$

However, since f is defined such that $f(\overline{y}_1, E) = 0$ for every $E \in \mathcal{A}$, we obtain that $a_A = 0$. This completes the proof of (12.5).

Clearly (12.5) implies that

$$\frac{f(\overline{y}, A)}{f(\overline{y}, X)} = \alpha_A \text{ for every } y \in Y \text{ for which } f(\overline{y}, X) \neq 0$$

Since the non-null set $A \in \mathcal{A}$ was arbitrary, we can now consider the function $\gamma : \mathcal{A} \to \mathbb{R}$ defined by

$$\gamma(E) = \begin{cases} \alpha_A \text{ if } A \in \mathcal{A} \text{ and } A \text{ is non-null in } (\mathcal{G}, \succsim) \\ 0 \text{ if } A \text{ is a null set in } (\mathcal{G}, \succsim) \end{cases} \tag{12.7}$$

Clearly (12.7) implies that γ is a measure on $\mathcal{A}$ and that γ is continuous with respect to μ.

Now we define the function $\overline{u} : Y \to \mathbb{R}$ by $\overline{u}(y) = f(\overline{y}, X)$ for every $y \in Y$, and we define the measurable functions $\alpha : X \to \mathbb{R}$ as an arbitrary Radon Nikodým derivative of the measure γ. Since $f(\cdot, X) : \mathcal{A} \to \mathbb{R}$ is continuous in t, $\overline{u}$ is continuous.

Now we just have to show that for any $g \in \mathcal{G}$ we have that

$$f(g, A) = \int_A \overline{u}(g(x))\alpha(x)d\mu$$

Indeed let for any $g \in \mathcal{G}$, g is not a constant function, $u_g : X \to \mathbb{R}$ be a Radon Nikodým derivative of the measure $f(g, \cdot) : \mathcal{A} \to \mathbb{R}$. Clearly for any $y \in Y$, $\overline{u}(y)\alpha(x)$ is a Radon Nikodým derivative of the measure $f(\overline{y}, \cdot) : \mathcal{A} \to \mathbb{R}$. Now we use lemma 55 to obtain that

$$u_g(x) = \lim_{y_n \to g(x)} \overline{u}(y_n)\alpha(x) \text{ a.e. for every } g \in \mathcal{G}.$$

Consequently, we have that

$$u_g(x) = \overline{u}(g(x))\alpha(x) \text{ a.e.}$$

and therefore for any $g \in \mathcal{G}$ and $A \in \mathcal{A}$

$$f(g, A) = \int_A \overline{u}(g(x))\alpha(x)d\mu$$

■

12.4 $f(g, A) = \int_A \overline{u}(t)e^{-\delta t}d\lambda$

We shall now consider a set $\mathcal{G}$ of **programs over time**, i.e. we let

$$\mathcal{G} \subset \{g : (T, \mathcal{B}, \lambda) \to Y | g \text{ is measurable}\}$$

where T is time, $T = [0, \infty[$, $\mathcal{B}$ the Lebesgue measurable subsets of T, and λ Lebesgue measure.

Let us assume that $(\mathcal{G}, \succsim)$ is an independent mixture. We shall now state what it means that $(\mathcal{G}, \succsim)$ is translation invariant. To do this let $k \in R_+$, and define $T_k = \{t \in T | t \geq k\}$. Now we define for each $g \in \mathcal{G}$ the function $g^k : T_k \to Y$ by

$$g^k(t) = g(t - k) \text{ for every } t \in T_k$$

If, for every $g \in \mathcal{G}$, $g^k \in \mathcal{G}|T_k$ then we can, corresponding to the preordering $\succsim$, define the preordering $\succsim_k$ on $\mathcal{G}$ by

$$g_1 \succsim_k g_2 \Leftrightarrow g_1^k \succsim_{\{t|t \geq k\}} g_2^k$$

Using the definition of $\succsim_{\{t|t\geq k\}}$we have that $g_1 \succsim_k g_2$ if and only if

$$(h|T_k^c, g_1^k) \succsim (h|T_k^c, g_2^k) \text{ for every } h \in \mathcal{G}.$$

We now define a translation invariant mixture

Definition 142 (translation invariant) *Let $(\mathcal{G}, \succsim)$ be an independent mixture of programs over time. Then $(\mathcal{G}, \succsim)$ is translation invariant if for every $g \in \mathcal{G}$ and $k \in \mathbb{R}_+, g^k \in \mathcal{G}|T_k$, and if for every $k \in \mathbb{R}_+$, the preordering $\succsim_k$ on $\mathcal{G}$ is identical with $\succsim$.*

That the independent mixture $(\mathcal{G}, \succsim)$ is translation invariant means that the relation between any two programs g_1 and g_2 does not change, if both programs are postponed identically.

Theorem 58 *Let Y be complete, connected, and separable metric space and $(T, \mathcal{B}, \lambda)$ a complete positive σ-finite measure space. Moreover, let $(\mathcal{G}, \succsim)$ be a translation invariant, independent, and connected mixture with more than two non-null sets containing the constant functions. Assume that $(\mathcal{G}, \succsim)$ is continuous with respect to λ and that the order topology is coarser than t.*

Then there exists a continuous function $\overline{u} : Y \to \mathbb{R}$ and $\delta > 0$ such that when the function

$$f : \mathcal{G} \times \mathcal{B} \to \mathbb{R}$$

is defined by

$$f(g, A) = \int_A \overline{u}(g(t))e^{-\delta t} d\lambda$$

then $f(\cdot, A)$ is strictly monotonic on $\mathcal{G}|A$ for all $A \in \mathcal{B}$.

Proof. Let $(y_n)_{n=1}^{\infty}$ be a dense subset of Y and let $f : \mathcal{G} \times \mathcal{B} \to$ be determined by theorem 26.

As the assumptions of theorem 52 are satisfied then there exists a measurable function $u : T \times Y \to \mathbb{R}$ such that

$$f(g, E) = \int_E u(t, g(t)) d\lambda \qquad \forall g \in \mathcal{G}, \forall E \in \mathcal{B}$$

We shall assume that u has been constructed as specified in the proof of theorem 52.

Now, let $k \in \mathbb{R}, k \geq 0$, and consider the preordering $\precsim_k$ on $\mathcal{G}$. By the definition of $\precsim_k$ we have that

$$g_1 \precsim_k g_2 \Longleftrightarrow \int_k^\infty u(t, g_1(t-k))d\lambda \leq \int_k^\infty u(t, g_2(t-k))d\lambda$$

Since $(\mathcal{G}, \precsim)$ is translation invariant we have that $\precsim_k$ is identical with $\precsim$. Consequently the function

$$f^k : \mathcal{G} \times \mathcal{B} \to \mathbb{R}$$
where
$$f^k(g,E) = \int\limits_{E+\{k\}} u(t,(t-k))d\lambda$$

is a function having the properties stated in the conclusion of theorem 26. Therefore there exists $\alpha(k) \in \mathbb{R}, \alpha(k) > 0$, and a measure μ_k such that

$$f^k(g,E) = \alpha(k)f(g,E) + \mu_k(E), \text{ for every } g \in \mathcal{G} \text{ and every } E \in \mathcal{B}$$

First we remark that $\mu_k(E) = 0$ for every $E \in \mathcal{B}$. This follows since f has been chosen such that $f(\overline{y}_1, E) = 0$ for all $E \in \mathcal{B}$. Consequently we obtain that

$$\int\limits_{E+\{k\}} u(t,g(t-k))d\lambda = \alpha(k) \int\limits_E u(t,g(t))d\lambda \tag{12.8}$$
for every $E \in \mathcal{B}$, every $g \in \mathcal{G}$, and every $k \in T$

We shall now prove that (12.8) implies that there exists a continuous function $\overline{u} : Y \to \mathbb{R}$ and $\delta > 0$ such that for a.e. $t \in T$

$$u(y,t) = \overline{u}(y)e^{-\delta t} \text{ for all } y \in Y$$

Let us define the function $H : T \times Y \to \mathbb{R}$ by

$$H(s,y) = \int_0^s u(t,y)d\lambda$$

(12.8) then implies that for every $s, t, k \in T$ and every $y \in Y$ we have

$$H(t+k, y) - H(s+k, y) = \alpha(k)(H(t, y) - H(s, y))$$

Letting $s = 0$, we obtain that

$$H(t+k, y) - \alpha(k)H(t, y) = H(k, y) \quad \text{for every } t, k \in T \text{ and every } y \in Y \tag{12.9}$$

By symmetry we also obtain from (12.9)

$$H(t+k, y) - \alpha(t)H(k, y) = H(t, y) \quad \text{for every } t, k \in T \text{ and every } y \in Y \tag{12.10}$$

and consequently

$$H(t, y)(1 - \alpha(k)) = H(k, y)(1 - \alpha(t)) \quad \text{for every } k, t \in T \text{ and every } y \in Y \tag{12.11}$$

We now show that there exists $k_0 \in T$ such that $\alpha(k_0) \neq 1$ and such that there exists $y_0 \in Y$ with $H(k_0, y_0) \neq 0$.

Indeed assume that such k_0 does not exist. It is easily seen that (12.9) and (12.10) then imply that

$$H(t+k, y) = H(t, y) + H(k, y) \text{ for every } t, k \in T \text{ and every } y \in Y$$

Since $H(\cdot, y) : T \to \mathbb{R}$ is continuous, it is well known (see e.g. (Aczél (1966) [5], p. 34) that this implies that $H(\cdot, y)$ is of the form

$$H(t, y) = c(y)t, \text{ where } c(y) \in \mathbb{R}$$

Now we remark that $c(y) = 0$ for every $y \in Y$, since if $c(y') \neq 0$ for some $y' \in Y$ then

$$f(\overline{y}', T) = \int_0^\infty u(y', t) d\lambda = +\infty$$

which is a contradiction. Consequently we obtain that for every $y \in Y, u(y, t) = 0$ for a.e. $t \in T$. However using the construction of u, this implies that for a.e. $t \in T, u(y,t) = 0$ for every $y \in Y$. However, then

$$f(g,E) = \int_E u(t,g(t))d\lambda = 0$$

for every $E \in \mathcal{B}$ and every $g \in \mathcal{G}$, which contradicts that $(\mathcal{G}, \precsim)$ contains more than non-null sets.

Consequently there exists $k_0 \in T$ such that $\alpha(k_0) \neq 1$ and such that for some $y_0 \in Y$ we have $H(k_0, y_0) \neq 0$.

Let us choose such $k_0 \in T$ and define $v : Y \to \mathbb{R}$ by

$$v(y) = \frac{H(k_0,y)}{1-\alpha(k_0)}$$

(12.11) then implies the following expression for H

$$\begin{array}{c} H(t,y) = (1-\alpha(t))v(y) \\ \text{for every } t \in T \text{ and every } y \in Y \end{array} \tag{12.12}$$

Using this, we obtain from (12.10) that

$$\begin{array}{c} v(y)(\alpha(t+k) - \alpha(t)\alpha(k)) = 0 \\ \text{for every } t,k \in T \text{ and every } y \in Y \end{array}$$

and consequently, since $v(y_0) \neq 0$ that

$$\alpha(t+k) = \alpha(t)\alpha(k) \qquad \text{for every } t,k \in T \tag{12.13}$$

Moreover we notice that $\alpha : T \to \mathbb{R}$ is continuous.

Indeed assume $k \in T$ and $\lim_n k_n = k$. By the definition of α we obtain

$$\lim(\alpha(k_n)\int_0^{k_0} u(t,y_0)d\lambda) = \lim \int_{k_n}^{k_0+k_n} u(t,y_0)d\lambda =$$
$$\int_k^{k_0+k} u(t,y_0)d\lambda = \alpha(k)\int_0^{k_0} u(t,y_0)d\lambda.$$

Since $\int_0^{k_0} u(y_0,t)d\lambda \neq 0$ we consequently have that $\lim_n \alpha(k_n) = \alpha(k)$.

Now it is well known (see e.g. (Aczél (1966) [5], p. 38) that a continuous function $\alpha : T \to \mathbb{R}$ which satisfies (12.13) is either of the form $\alpha(t) = 0$ for every $t \in T$ or $\alpha(t) = e^{ct}$, where $c \in \mathbb{R}$. Since $\alpha(0) \neq 0$ we consequently get the existence $\delta \in \mathbb{R}$ such that

$$\alpha(t) = e^{-\delta t} \text{ for every } t \in T$$

From (12.11) we now get that

$$\int_0^t u(s,y)d\lambda = v(y)(e^{-\delta t} - 1) \text{ for every } t \in T, \text{ and every } y \in Y$$

As above we obtain $\delta \neq 0$ and consequently for every $y \in Y$ we have that

$$u(y,t) = \frac{v(y)}{-\delta} e^{-\delta t} \text{ for a.e. } t \in T \tag{12.14}$$

As

$$|f(\overline{y}, T)| = |\int_0^\infty \frac{v(y)}{-\delta} e^{-\delta t} d\lambda| < +\infty$$

we have moreover that $\delta > 0$.

Now let $\overline{u} : Y \to \mathbb{R}$ be defined by

$$\overline{u}(y) = \frac{v(y)}{-\delta}$$

Since

$$\overline{u}(y) = \frac{f(\overline{y}, [0, k_0])}{(1 - e^{-\delta k_0})\delta}$$

and the mapping $y \to f(\overline{y}, [0, k_0])$ is continuous, then $\overline{u}$ is continuous.

Finally we just have to prove that (12.14) can be sharpened to:

$$\text{for a.e. } t \in T, u(t,y) = \overline{u}(y)e^{-\delta t} \text{ for every } y \in Y$$

However this follows easily from the construction of $u : T \times Y \to \mathbb{R}$ and the continuity of $\overline{u}$. Indeed

$$u : T \times Y \to \mathbb{R}$$
is defined by
$$u(t,y) = \limsup_{y_n \to y} u_{\overline{y}_n}(t)$$

where $u_{\overline{y}_n}(\cdot)$ is an arbitrary Radon Nikodým derivative with respect to λ of the measure $f(\overline{y}_n, \cdot) : \mathcal{B} \to \mathbb{R}$. Since for every $n \in N, u_{\overline{y}_n}(t) = u(t, y_n)$, a.e. $t \in T$, we have for a.e. $t \in T$,

$$u(t, y) = \limsup_{y_n \to y} u(t, y_n) \text{ for every } y \in Y.$$

Now let

$$T_0 = \{t \in T | u(t, y) = \limsup_{y_n \to y} u(t, y_n) \text{ for every} y \in Y\}$$

and let for $n \in N$,

$$N_n = \{t \in T | u(t, y_n) \neq \overline{u}(y_n) e^{-\delta t}\}$$

Assume $t \in T_0 \backslash \bigcup_{n=1}^{\infty} N_n$ and $y \in Y$. Then, since $\overline{u}$ is continuous, we have

$$u(t, y) = \limsup_{y_n \to y} \overline{u}(y_n) e^{-\delta t} = \overline{u}(y) e^{-\delta t}$$

Consequently for every $t \in T_0 \backslash \bigcup_{n=1}^{\infty} N_n$

$$u(t, y) = \overline{u}(y) e^{-\delta t} \text{ for all } y \in Y$$

Since

$$\lambda(T_0^c \cup \bigcup_{n=1}^{\infty} N_n) = 0$$

this implies that for every $E \in \mathcal{B}$ and $g \in \mathcal{G}$ we have that

$$f(g, E) = \int_E \overline{u}(g(t)) e^{-\delta t} d\lambda.$$

Since $f(\cdot, E)$ is strictly monotonic on $\mathcal{G}|E$, the proof is complete ■

12.5 Notes

1. As in chapter 11 the theorems in this chapter have two classical interpretations in economic theory. Namely $X = [0, \infty[$

is time, or $(X, \mathcal{A}, \mu)$ is the measure space of events. In both cases, theorem 56 states conditions on $(\mathcal{G}, \succsim)$ which allow us to define a utility function $\overline{u}$ on Y and a generalized time preference function (state of nature preference function) $\beta : X \times \overline{u}(Y) \to \mathbb{R}$ such that a utility representation of $(\mathcal{G}, \succsim)$ is obtained by

$$U(g) = \int_X \beta(x, \overline{u}(g(x)) d\mu \qquad (12.15)$$

The representation (12.15) of $(\mathcal{G}, \succsim)$ allows us to talk about a fixed preordering on Y, namely the one induced by $\overline{u}$.

However the utility representation which at a given $x \in X$ should be used of this preordering in order to get the integral representation of $(\mathcal{G}, \succsim)$ is the representation $\beta(x, \overline{u}(\cdot)) : Y \to \mathbb{R}$.

The function β is a generalized time preference function (state of nature preference function) since β not only depends on x but also on the utility level r, measured in $\overline{u}$, obtained at x.

2. The representation of $(\mathcal{G}, \succsim)$ obtained in theorem 57 is the one which is often used as a starting point in economic theory when $X = [0, \infty[$. There is, as in theorem 56, a fixed preordering on Y and a time preference function $\alpha : X \to \mathbb{R}$.

3. In theorem 58 we assume that $X = [0, \infty[$ and give conditions so that the time preference function has the special form $\alpha(t) = e^{-\delta t}$ i.e., where the utility of a program is the integral of the discounted utilities (measured in $\overline{u}$) with a discount factor $\delta > 0$.

   For the case of discrete time this last problem has previously been studied by e.g. Koopmans (1960) [114], Koopmans, Diamond, Williamson (1964) [62], Koopmans (1972) [116] and Easton (1972) [65]. E.g. in [116] are stated conditions on $(\mathcal{G}, \succsim)$ which imply that there exists a utility function u on the choice space Y and a discount factor $0 < \delta < 1$ such $\succsim$ on the programs can be represented by the function

$$U(x_1, x_2, \dots, x_t, \dots) = \sum_{t=1}^{\infty} \delta^{t-1} u(x_t).$$

Results about representation theorems for preferences over time in case the basic relation is not total can be found in chapter 17 page 231.

Part V

Decompositions and Uncertainty

13

Decompositions. Uncertainty

13.1 Introduction

Agents in an economy, a game, or any other social system make decisions or choose between alternatives. Theories about such decisions or choices often include as an assumption that agents have preferences in the form of a relation on a set of alternatives. A very used special case is that the set of alternatives is a set of functions, that the preferences can be represented by real numbers, and can be decomposed into utility functions and measures, where the preference can be expressed as the expected value of the utility function with respect to the measure.

In this way a distinction between utility in the form of an utility function and knowledge in the form of a measure is introduced.

This again means that theories about how preferences change with observations can be based on conditional probabilities i.e. only the measures have to change, the utility may remain the same.

The theory also has the extremely convenient property that the mathematics of combining probabilities and integrals on factors in a product and probabilities and integrals on the product itself is well understood.

By introducing the idea of a parameter space the same representation theorems can be used to give a foundation for statistics. See Savage (1954) [160] and chapter 18 page 243.

This theory and its applications have been criticized for lack of realism (see Allais (1952, 1953) [8, 9], Kahneman Tversky (1979) [101] and many others), for not formulating a concept of uncertainty different from risk (see Bewley (1986,1987,1988,2001) [21, 22, 23, 24], Ellsberg (1961) [68], Keynes (1921) [106], and Knight

(1921)[1] [110]), and for making the questionable assumption that a total order and a probability measure exist on subsets of a parameter space. (See Lehmann (1990) [121] and many others)

Many attempts have been made to formulate more satisfactory theories. For most of the attempts no satisfactory theory of how to go to conditional knowledge and preference and how to combine knowledge and preference on factors exist.

One of the main purposes of this book is to formulate an alternative theory, which - at least from a mathematical point of view - is satisfactory. The first step in such a theory is to generalize the classical representation theorems for totally preordered function spaces to function spaces with just a relation not assumed to be total or even transitive. In this way one obtains precisely defined concepts of uncertainty. This first step is taken in this chapter.

The framework is the same as in chapters 10 and 9. X, Y and Z are arbitrary sets and $\mathcal{A}$ a system of subsets of X. $\mathcal{G}$ $(\mathcal{H})$ is a space of functions defined on X with values in Y (Z). In chapter 10 also a probability measure α on X is given. Finally $graph\mathcal{P} = Q \subset \mathcal{G} \times \mathcal{H}$

It is in this chapter assumed that $Y = Z$ and $\mathcal{G} = \mathcal{H}$. The representations from chapters 10 and 9 are the starting points for this chapter.

13.2 von Neumann Morgenstern preferences

Theorem 48 page 139 gives the existence of a function $f : \mathcal{G} \times \mathcal{G} \to \mathbb{R}$ representing a $\mathcal{P}$ and an affine function $\widetilde{f}$ on $\chi(\mathcal{G} \times \mathcal{G})$ (where $\chi(g, g')$ is the joint distribution of (g, g')) such that

$$\widetilde{f}(\chi(g, g')) = f(g, g') > 0 \Leftrightarrow g \in \mathcal{P}(h)$$

Definition 143 (pure preference) *The pure preference for g over g' is expressed by*

$$p : \mathcal{G} \times \mathcal{G} \to \mathbb{R}$$
$$\textit{where}$$
$$p(g, g') = \tfrac{1}{2}(f(g, g') - f(g', g))$$

[1] See note 13.4.1

Definition 144 (uncertainty) *The uncertainty in the choice between g and g' is expressed by*

$$q : \mathcal{G} \times \mathcal{G} \to \mathbb{R}$$
where
$$q(g, g') = -\tfrac{1}{2}\left(f(g, g') + f(g', g)\right)$$

The basic representation - with this rewriting - now takes the form

$$g \in \mathcal{P}(g') \Leftrightarrow p(g, g') > q(g, g')$$

The pure preference is skew symmetric and the uncertainty is symmetric. Both functions are equal to affine functions on the space of distributions.

If f has an integral representation

$$g \in \mathcal{P}(g') \Leftrightarrow \int w d\lambda > 0$$

(see chapter 11), one function g is preferred to another g', if the expected value of a preference function is larger than 0. This can be rewritten as one function g is preferred to another g', if the expected value of a pure preference function is larger than the expected value of an uncetainty function

Definition 145 (pure preference) *The pure preference for y over y' is expressed by*

$$u : Y \times Y \to \mathbb{R}$$
where
$$u(y, y') = \tfrac{1}{2}\left(w(y, y') - w(y', y)\right)$$

The uncertainty in the choice between y and y' is expressed by

$$v : Y \times Y \to \mathbb{R}$$
where
$$v(y, y') = -\tfrac{1}{2}\left(w(y, y') + w(y', y)\right)$$

The integral form of the representation now takes the form

$$g \in \mathcal{P}(g') \Leftrightarrow \int u d\lambda > \int v d\lambda$$

So one function g is preferred to another function g' if and only if the expected value of the preference function is larger than the expected value of the uncertainty. Both expectations are with respect to the joint distribution of the two functions. Both pure preference and uncertainty are real functions defined on $Y \times Y$.

Example 11 *(example 10 page 143 continued) The uncertainty in the choice between two portfolios x and y is $V(x,y) = V(y,x) = -\frac{1}{2}(W(y,x) + W(x,y))$ in this illustration the uncertainty is*

$$V(y,x) = \frac{1}{2}(\gamma_2 - \gamma_1)(x-y)B(x-y)$$

i.e. proportional to the second moment of the random variable obtained from $(x-y)$. This means for example that two portfolios with the same expected value and the same variance cannot be compared. Both $W(y,x)$ and $W(x,y)$ will be negative (unless the variance of the difference $x-y$ is 0).

13.3 Function spaces

13.3.1 $Y = Z = \{0,1\}$. Subjective probabilities and uncertainty

It is convenient to start with uncertainty spaces and to repeat the result from section 9.5 page 118

Theorem 59 (Theorem 37) *Let $(X, \mathcal{A}, \mathcal{P})$ be an uncertainty space, where $(X, \mathcal{A})$ has more than three disjoint subsets A_i for which $A_i \in \mathcal{P}(\emptyset)$. Let assumption A hold, let $\mathcal{P}$ be independent and open, and let $\mathcal{A}|A_i$ be connected. Then there exist additive functions*

$$\alpha : \mathcal{A} \to \mathbb{R}_+ \text{ and } \beta : \mathcal{A} \to \mathbb{R}_+$$
$$\text{where}$$
$$\alpha(A \backslash B) - \beta(B \backslash A) > 0 \Leftrightarrow A \in \mathcal{P}(B)$$

Rewriting we get

$$\lambda(A) - \lambda(B) > \mu(A \triangle B) \Leftrightarrow A \in \mathcal{P}(B)$$
$$\text{where}$$
$$\alpha = \lambda - \mu,\ \beta = \lambda + \mu$$

where λ can be regarded as a probability measure and μ as an uncertainty measure. A is then more "likely" than B if the difference in probability is larger than the uncertainty. The uncertainty is measured on the symmetric difference of the two sets.

Two other ways of rewriting the representation theorem for subjective probability and uncertainty will be useful

$$\begin{gathered} A \in \mathcal{P}(B) \\ \Leftrightarrow \\ \lambda(A) - \lambda(B) > 0, \text{ for } \forall\lambda \text{ with } \alpha \le \lambda \le \beta \\ \Leftrightarrow \\ \lambda(A) - \lambda(B) > 0, \text{ for } \forall\lambda \text{ with } \lambda(X) = 1 \\ \text{and } p(A) \le \lambda(A) \le b(A) \\ \text{where} \\ p(A) = \frac{\alpha(A)}{\alpha(A)+\beta(A^c)} \text{ and } b(A) = \frac{\beta(A)}{\beta(A)+\alpha(A^c)} \end{gathered} \tag{13.1}$$

$p, b : \mathcal{A} \to [0,1]$ are non-additive set functions, but have of course very special properties[2], they can be called lower and upper probabilities. A is then more "likely" than B if the probability of A is larger than the probability of B for all probability measures between the lower and upper "probabilities". There will be large advantages in reserving the word probability for additive set functions, so this terminology will not be used here. All the different reformulations of the basic theorem will however be useful.

13.3.2 $X = \{1, 2, \cdots, n\}$ $\left(\prod_{i \in X} Y_i, \mathcal{P}\right)$

For $X = \{1, 2, \cdots, n\}$ the notation $(y_1, y_2, \cdots, y_n)$ is used for the values of the functions g.

The representation in corollary 15 page 118 is

$$\begin{gathered} f : (Y_1 \times Y_2 \times \cdots \times Y_n) \times (Y_1 \times Y_2 \times \cdots \times Y_n) \to \mathbb{R} \\ \text{such that} \\ f(y_1, y_2, \cdots, y_n, y_1', y_2', \cdots, y_n') = \sum w_i(y_i, y_i') \\ \text{and} \\ f(y, y') > 0 \Leftrightarrow y \in \mathcal{P}(y') \end{gathered}$$

---

[2] $p(A) + b(A^c) = 1$. In the terminology of mathematical programming $-p$ and b are fully submodular and compliant. See Frank Tardos (1988) [79].

f can trivially be rewritten

$$f(y,y') = \sum w_i(y_i,y_i') = \sum u_i(y_i,y_i') - \sum v_i(y_i,y_i')$$

where u_i is skew symmetric and v_i is symmetric expressing pure preference and uncertainty, $w_i = u_i - v_i$

Definition 146 (pure preference) *The pure preference for y_i over y_i' is expressed by*

$$\begin{gathered} u_i : Y \times Y \to \mathbb{R} \\ \textit{where} \\ u_i(y,y') = \tfrac{1}{2}\left(w_i(y_i,y_i') - w_i(y_i',y_i)\right) \end{gathered}$$

Definition 147 *The uncertainty in the choice between y_i and y_i' is expressed by*

$$\begin{gathered} v_i : Y \times Y \to \mathbb{R} \\ \textit{where} \\ v_i(y,y') = -\tfrac{1}{2}\left(w_i(y,y') + w_i(y',y)\right) \end{gathered}$$

13.3.3 Y and X general

The starting point is the representation in subsection 11.2 page 152

$$f(g,h,A) = \int_A W(g(x),h(x),x)\,d\chi(g,h)$$

Chose functions g_0 and g_1 such that $f(g_1,g_0,A) > 0$ for all A for which such functions exist. Define

$$\begin{gathered} \alpha : \mathcal{A} \to \mathbb{R} \text{ by } \alpha(\cdot) = f(g_1,g_0,\cdot) \\ \text{and} \\ \beta : \mathcal{A} \to \mathbb{R} \text{ by } \beta(\cdot) = -f(g_0,g_1,\cdot) \end{gathered}$$

Then some trivial manipulation gives

$$f(g,h,A) > 0 \Leftrightarrow \int_A w(g(x),h(x),x)\,d\gamma > 0 \text{ for } \alpha \leq \gamma \leq \beta$$

$$\begin{array}{ll} f(g,h,A) = & \\ \int_A u(g(x),h(x),x)\,d\lambda & \text{(Expected pure preference} \\ -\int_A v(g(x),h(x),x)\,d\lambda & \text{- expected } Y\text{-uncertainty} \\ -\int_A |w(g(x),h(x),x)|\,d\mu & \text{- uncertainty on } \mathcal{A}) \end{array}$$

where

$$\begin{array}{ll} w, u, v : Y^2 \times X \to \mathbb{R}, & \\ w = u - v = w^+ + w^-; w^+, -w^- \geqq 0 & \\ u(\cdot, \cdot, x) \text{ is skew symmetric.} & \text{(Pure preference)} \\ v(\cdot, \cdot, x) \text{ is symmetric.} & \text{(Uncertainty on } Y\text{)} \\ w^+ = \frac{d\chi(g,h)}{d\alpha} W^+,\ w^- = \frac{d\chi(g,h)}{d\beta} W^- & \\ \text{and as before} & \\ \alpha = \lambda - \mu,\ \beta = \lambda + \mu & \end{array}$$

A function g is now preferred to another function h if the expected pure preference - $\int_X u d\lambda$ - is larger than the sum of the two types of uncertainty. First the expected value - $\int_X v d\lambda$ - of the uncertainty on Y with respect to the same probability measure, and secondly the uncertainty on $\mathcal{A}$ weighted by the absolute value of the preference - $\int_X |w| \, d\mu$.

13.4 Historical notes

The results in this chapter are new, but of course trivial rewritings of the earlier representation theorems. The concept of uncertainty being formalized here has been called Knightian uncertainty.

13.4.1 Knight

LeRoy and Singel (1987) [124] argue convincingly that Knight (1921) [110] wanted to distinguish between probabilities 1) known from symmetry (coins or dice), 2) known from many observations (mortality etc.), or 3) subjective probabilities (probabilities for single events). The subjective probabilities could be different for different people, and one persons subjective probability would be unknown to other persons. All the probabilities were however probabilities and thus representations of total preorders. If that view is correct Knight did not invent Knightian uncertainty.

13.4.2 Keynes

There are not the same doubts about Keynes (1921) [106], Keynes kept his views also in Keynes (1936) [107]:

> It would be foolish, in forming our expectations, to attach great weight to matters which are very uncertain. (Footnote): By very uncertain I do not mean the same thing as "very improbable"

(Keynes (1936) [107] page 148)
Both Keynes and Knight insisted that "their" uncertainty could not be formalized.

13.4.3 von Neumann Morgenstern

The position about the possibilities for formalizing "Knightian" uncertainty of von Neumann and Morgenstern was:

> one may question the postulate of axiom (3:A:a) for $u > v$, i.e. the completeness of this ordering. (Page 19)
>
> The axiom (3:A:a) expresses the completeness of the ordering of all utilities, I.e. the completeness of the individuals's system of preferences It is very dubious, whether the idealization of reality which treats this postulate as a valid one, is appropriate or even convenient (page 630).
>
> Let us consider this point for a moment. We have conceded that one may doubt whether a person can always decide which of two alternatives – with the utilities u, v – he prefers.
>
> If the general comparability assumption is not made, a mathematical theory – based on $\alpha u + (1 - \alpha) v$ together with what remains of $u > v$ – is still possible. It leads to what may be described as a many-dimensional vector concept of utility. This is a more complicated and less satisfactory set-up, but we do not propose to treat it systematically at this time. (page 19-20)

(all quotes from von Neumann Morgenstern (1944) [141]).

These ideas has been taken up by Shapley Baucells (1998) [165] (based on an earlier paper by Shapley, probably written at the same time as Aumann (1962) [11]).

von Neumann and Morgenstern took as an "axiom" that some concept of utility should appear in the representation. In the representations in this book the utility concept has been replaced by preference, and utility is a special case where the preference of A over B is the difference in utility.

13.4.4 Savage

The position of Savage was:

> There is some temptation to explore the possibilities of analyzing preferences among act as a partial ordering, that is, in effect to replace part 1 [Part 1. Either $x \leq y$ or $y \leq x$. Savage (1954) [160] page 17] of the definition of simple orderings by the very weak proposition $f \leq f$, admitting that some pairs of acts are incomparable. This would seem to give expression to introspective sensation of indecision or vacillation, which we may be reluctant to identify with indifference. My own conjecture is that it would prove a blind alley losing much in power and advancing little, if at all, in realism; but only an enthusiastic exploration could shed light on the question.

(Savage (1954) [160] page 21*)*.

13.4.5 Aumann

Aumann argued for incomplete preferences in 1962:

> We are concerned here another of the axioms, *the completeness axiom.* This axiom says that *given any pair of lotteries, the individual either prefers one to the other or is indifferent between them.* It specifically excludes the possibility that an individual may be willing and able to arrive at preference decisions only for certain pairs of lotteries, while for others he may be unwilling or unable to at a decision; ... Of all the axioms of utility theory, the completeness axiom is perhaps the most questionable

(Aumann (1962) [11] page 446). Like von Neumann Morgenstern, Aumann could not get away from utility and was thus unable to get nice representation theorems.

13.4.6 Friedman

> In his seminal work, Frank Knight drew a sharp distinction between *risk,* as referring to events subject to a known or knowable probability distribution and *uncertainty,* as referring to events for which it was not possible to specify numerical probabilities. I have not referred to this distinction because I do not believe it is valid. I follow L.J. Savage in his view of *personal probability,* which denies any distinction along these lines. We may treat people as if they assigned numerical probabilities to every conceivable event.

(Friedman (1976) [80] page 282). This position has probably been the majority opinion among economists.

13.4.7 Bewley

> A theory of choice under uncertainty is proposed which removes the completeness assumption from the Anscombe-Aumann formulation of Savage's theory and introduces an inertia assumption. The inertia assumption is that there is such a thing as the status quo and an alternative is accepted only if it is preferred to the status quo. This theory is one way of giving rigorous expression to Frank Knight's distinction between risk and uncertainty.

(Bewley (1986) [21] abstract).

> I modify the usual axiomatic basis of Bayesian decision theory by eliminating the assumption that preferences over lotteries are complete and by adding an inertia hypothesis.

(Bewley (2001) [24] page 43).
The results in this book generalizes Bewley's results to the general Savage and von Neumann Morgenstern framework. Bewley's ideas about inertia are probably very important for applications of the results from this book - in particular for the problems in chapter 17 but they are not taken up in this book.

13.5 Conclusion

Expected utility theory and the special case of subjective probability are based on assumptions for representation theorems. An order relation is assumed to be total, transitive and to satisfy convexity or independence conditions. A concept of utility shows up in these representation theorems, but utility of course does not exist before the representation theorems. The contribution of this chapter is to show that by giving up the assumption that the relation is total and transitive (but keeping the other assumptions) concepts of uncertainty appear as natural parts of the representation theorems. So the belief that uncertainty could not be formalized has turned out to be incorrect. Uncertainty can be represented by real numbers - just like preferences and probabilities. This uncertainty measures how far a given relation is from being a total relation. What had to be given up turned out to be utility and not the additivity of the representations.

14

Uncertainty on products

14.1 Introduction

For many applications the space of functions has not one but several preference relations expressing the knowledge and uncertainty. The set X may be a product space with separate uncertainty on each of the sets.

For example the profit of a firm, the future income of a household, the future prices of goods, bonds or stocks etc. will be functions of more than one random or uncertain variable. In a game or a social system the outcome for one player or agent will depend on the strategies of the other agents and possibly on some other variables. All of these may have separate uncertainty. The way agents combine uncertainty on several factors or on several sub $\sigma-$algebras is very important for most applications of an uncertainty concept.

Any foundation of statistics will involve a product a parameter space and a space of outcomes, so also in this application of an uncertainty concept the theory developed so far can not be enough.

The one dimensional theory presented so far in this book can therefore at most be the starting point for a theory of uncertainty with any applicability.

The classical case where preferences are represented by measures, utility functions, and integrals is of course well understood.

1. A measure on a set determines measures on non-null subsets. In general measures on a set with a $\sigma-$algebra may determine conditional measures on sub $\sigma-$algebras.
   When measures are regarded as order preserving real function, we get that a preorder on a system of subsets determines preorders on non-null subsets. In general preorders

on a $\sigma-$algebra may determine conditional preorders on sub $\sigma-$algebras. In applications these conditional preorders will be a natural first choice as conditional preorders given for example observations or in general given a sub $\sigma-$algebra.

2. Measures on products determine measures on the factors. Measures on the factors determine a product measure on the product, and determine a class of measures on the product with the given set of measures as marginal measures (Strassen (1965) [171]).
   When measures are regarded as order preserving real function, we get that a preorder on a system of subsets of a product determines preorders on subsets of the factors. Preorders on the factors determine a "product" preorder and class of preorders on subsets of the product, each of those measures on the product in return determines the original preorder on the product.

3. Integrals of functions defined on products can be taken in any order and are identical to the integral with respect to the product measure (Fubini). See for example Halmos (1961) [93] p. 148.
   When measures are regarded as order preserving real function, we get that a preorder on a set of functions on a product in two ways can be represented as a preorder on functions with values in a function space with a conditional preorder.

This nice theory does generalize in at least two forms, and the two different generalizations are both important for applications. The purpose of the following sections is to introduce these two possible generalizations.

In section 14.2 the relation between relations on subsets of products and relations on subsets on the factors is studied first. Then the relation between relations on functions on products and relations on functions on the factors is studied. Relation on functions or on subsets of factors can be trivially extended by monotonicity. The important result is however theorem 60. An extension of individual relations on subsets on factors or on sub $\sigma-$algebras yields a combined relation on subsets of the product or on the coarsest $\sigma-$algebra containing the individual sub $\sigma-$algebras. The combined relation is characterized using a very special convex set of probability measures on the product. Example 13 shows that the combined relation has properties not contained in the union of the individual relations.

Section 14.3 is a continuation of sections 8.9, 8.10 and 9.7. The starting point in these sections were relations on spaces of functions where the range also had a preorder and mean groupoid structure. A generalization of Fubini's theorem is studied. A relation on functions on a product of two factors can be represented by an integral with respect to one of the factors of conditional representations on the other factor - just like Fubini - but the independence condition has to be relaxed and the conditions are not symmetric in the two factors, so the "independence of the order" aspect of Fubini disappears.

This gives two level uncertainty and will be the starting point for the foundation of statistics sketched in chapter 18.

14.2 One level uncertainty on factors and products

14.2.1 $Y = Z = \{0,1\}, X = X_1 \times X_2$

The case $Y = Z = \{0,1\}, X = X_1 \times X_2$ will be treated first.

An extreme and simple example will illustrate what may happen

Example 12 *Let* $X_1 = X_2 = \{0,1\}$ *so the sample space is*

$$(0,0), (0,1), (1,0), (1,1)$$

and the probability of

$$[(0,0), (0,1)], [(0,0), (1,0)], [(0,1), (1,1)]$$
and
$$[(1,0), (1,1)]$$

are all $\frac{1}{2}$, *so there is no uncertainty on* X_1 *or* X_2. *We may anyway want complete uncertainty about the joint distribution, so the probability* ***and*** *the uncertainty of*

$$(0,0), (0,1), (1,0), \text{ and } (1,1)$$

are all $\frac{1}{4}$.

If we want to allow for uncertainty on the product without uncertainty on the factors - uncertainty only about the joint distribution - or if we just want to allow for more uncertainty the finer the algebra is, we obviously have to give up the close relation between measures on products and factors.

The form of representation in subsection 13.3.1 page 193 can be used to formulate uncertainty on products. Let $(X_1, \mathcal{A}^1), (X_2, \mathcal{A}^2)$ and $(X, \mathcal{A}_0)$ be measurable spaces relations where $X = X_1 \times X_2$ and $\mathcal{A}_0$ the product algebra, define

$$\mathcal{A}_1 = \{A \,|A = A_1 \times X_2, \; A_1 \in \mathcal{A}^1\}$$

$$\mathcal{A}_2 = \{A \,|A = X_1 \times A_2, \; A^2 \in \mathcal{A}^2\}$$

and note that

$$\mathcal{A}_1 \cap \mathcal{A}_2 = \{\emptyset, X\}$$

Let $\mathcal{P}_1, \mathcal{P}_2$ and $\mathcal{P}_0$ be relations on $(X_1, \mathcal{A}_1), (X_2, \mathcal{A}_2)$ and $(X, \mathcal{A}_0)$ respectively. $\mathcal{P}_1, (\mathcal{P}_2)$ will be considered as relations both on $\mathcal{A}^1$ and $\mathcal{A}_1$ (and on $\mathcal{A}^2$ and $\mathcal{A}_2$))

Definition 148 (Consistency) *Consistency between $\mathcal{P}_1, \mathcal{P}_2$ and $\mathcal{P}_0$ will be defined to mean that*

$$\begin{array}{l} A \in \mathcal{P}_i(B) \Longrightarrow A \in \mathcal{P}_0(B), (i = 1, 2) \\ A \notin \mathcal{P}_0(B) \Longrightarrow A \notin \mathcal{P}_i(B), (i = 1, 2) \end{array}$$

whenever $A, B \in \mathcal{A}_i$.

Remark 53 *Consistency between two relations is a very natural condition and it will be made in the following. It simply means that the uncertainty is largest on the finest algebra. If it does not hold the preference can simply be changed on the coarse algebra to the finer relation on the fine algebra.*

If the assumptions for the representation theorems (see theorem 37 page 119) hold we will for $A, B \in \mathcal{A}_i$, $(i = 0, 1, 2)$ have

$$A \in \mathcal{P}_i(B) \Leftrightarrow \alpha_i(A \setminus B) > \beta_i(B \setminus A)$$

or (see 13.3.1)

$$A \in \mathcal{P}_i(B) \Leftrightarrow \lambda_i(A) > \lambda_i(B) \text{ for all } \lambda_i \text{ with } \alpha_i \leqq \lambda_i \leqq \beta_i.\ \lambda_i : \mathcal{A}_i \to R_+$$

or

$$A \in \mathcal{P}_i(B) \Leftrightarrow \lambda_i(A) > \lambda_i(B) \text{ for all } \lambda_i \text{ with } p_i \leqq \lambda_i \leqq b_i.\ \lambda_i : \mathcal{A}_i \to [0,1]$$

The relations $\mathcal{P}_1, \mathcal{P}_2$ and $\mathcal{P}_0$ makes it possible to compare sets from $\mathcal{A}_1, \mathcal{A}_2$ and from $\mathcal{A}_0$. But adding further assumptions one can combine the relations to a relation defined on larger systems of sets - for example on $\mathcal{A}$ the coarsest $\sigma-$ algebra containing $(\mathcal{A}_i)_{i \in \{0,1,2\}}$ - expressing the combined preference. One example is to use the relations $\mathcal{P}_1$ and $\mathcal{P}_2$ on sets not in $\mathcal{A}_1$ or $\mathcal{A}_2$.

Definition 149 (Monotonicity) *$\mathcal{P}$ is monotone if*

$$A \in \mathcal{P}(B) \Rightarrow A_0 \in \mathcal{P}(B_0) \text{ for } A \subset A_0, B_0 \subset B$$

or equivalently

$$A \notin \mathcal{P}(B) \Rightarrow A_0 \notin \mathcal{P}(B_0) \text{ for } A_0 \subset A, B \subset B_0$$

Definition 150 (Monotonic extension) *$\mathcal{P}_m$ is the monotonic extension of $\mathcal{P}$ if $\mathcal{P}_m$ is the smallest monotone relation defined on $\mathcal{A}$ containing $\mathcal{P}$*[1]

Defining the combined relation $\mathcal{P}_m$ by assuming that it is the monotonic extension of $\mathcal{P}_i$ $(i = 0, 1, 2)$ we have in addition for $A, B \in \mathcal{A}$

$$A \in \mathcal{P}_m(B) \iff \begin{cases} \alpha(A \setminus B) > \beta(B \setminus A), A, B \in \mathcal{A} \\ \text{or} \\ \alpha_i(A_0 \setminus B_0) > \beta_i(B_0 \setminus A_0) \end{cases}$$

for some A_0, B_0 such that

$$A \supset A_0 \in \mathcal{A}_i \ni B_0 \supset B \text{ for } i = 1 \text{ or } 2$$

[1] $\mathcal{P}_0$ contains $\mathcal{P}_i$ if $graph\mathcal{P}_0 \supset graph\mathcal{P}_i$

Another extension will be called **the** extension and denoted $\mathcal{P}_c$. It is for $I = \{0,1,2\}$ defined on $\mathcal{A}$ by

Definition 151 ($\mathcal{P}_c$) *$\mathcal{P}_c$ is **the** extension of $\mathcal{P}_i$ $(i \in I)$*

$$A \in \mathcal{P}_c(B) \Longleftrightarrow \lambda(A) > \lambda(B) \text{ for all } \lambda \text{ with}$$
$$p_i \leqq \lambda|\mathcal{A}_i \leqq b_i.\ \lambda : \mathcal{A} \to [0,1]$$
$$\text{for all } i \in I$$

or equivalently

$$A \in \mathcal{P}_c(B) \Longleftrightarrow \lambda(A) > \lambda(B) \text{ for all } \lambda \text{ with}$$
$$p_i(A) \leqq \lambda(A) \leqq b_i(A).\ A \in \mathcal{A}_i \text{ for all } i \in I$$

Remark 54 *This extension assumes that the combined preferences can be expressed as restrictions on measures on $\mathcal{A}$ - the coarsest $\sigma-$algebra finer than $(\mathcal{A}_i)_{i=0,1,2}$.*

The two extensions will be defined in general in the next subsection. An example (example 13, page 211) will show that $\mathcal{P}_c$ can be much finer than just the union of the individual relations.

Making assumptions corresponding to stochastic independence and product measures further similar extensions can be obtained.

Conjecture 4 *An obvious conjecture is that a natural generalization*

$$A_1 \in \mathcal{P}_1(B_1) \text{ and } A_2 \in \mathcal{P}_2(B_2) \Longrightarrow A_1 \times A_2 \in \mathcal{P}_0(B_1 \times B_2)$$
$$A_1 \notin \mathcal{P}_1(B_1) \text{ and } A_2 \notin \mathcal{P}_2(B_2) \Longrightarrow A_1 \times A_2 \notin \mathcal{P}_0(B_1 \times B_2)$$

will imply product measures as the extension.

14.2.2 $Y = Z = \{0,1\}, (X, \mathcal{A}_i)_{i \in I}$

This generalizes without large changes to the case where there is a set of sub $\sigma-$algebras $(\mathcal{A}_i)_{i\in I}$ $(i \in I)$ of X, and relations $(\mathcal{P}_i)_{i\in I}$ on these $\sigma-$algebras. (X does not have to be a product). If

$$(\mathcal{P}_i)_{i\in I} \text{ are relations on } \mathcal{A}_i$$

and the conditions for the representation theorems hold, then we have representations $\forall i \in I$ of the form

$$\begin{gathered} A \in \mathcal{P}_i(B) \iff \lambda_i(A) > \lambda_i(B) \\ \text{for all } \lambda_i \text{ with} \\ \alpha_i \leqq \lambda_i \leqq \beta_i, \lambda_i : \mathcal{A}_i \to R \\ \alpha_i, \beta_i, \lambda_i \text{ additive,} \\ \text{or for all additive } \lambda_i \text{ with} \\ p_i \leqq \lambda_i \leqq b_i. \ \lambda_i : \mathcal{A}_i \to [0,1] \end{gathered}$$

Several special cases will be of interest. The σ-algebras $(\mathcal{A}_i)$ have of course a partial order $\subset$ and a special case will be that $\mathcal{A}_i \subset \mathcal{A}_j$, another special case will be that two σ-algebras are orthogonal in the sense that the largest σ-algebra contained in both is the trivial algebra $\{X, \emptyset\}$. The consistency requirements are as before

Definition 152 *Consistency for $(\mathcal{P}_i)_{i \in I}$ will be defined to mean that for any A, B where $\sigma(A,B) = \sigma(i)$ for some $i \in I$ such that $A, B \in \sigma(A,B)$ and*

$$\begin{gathered} A \in \mathcal{P}_i(B) \Longrightarrow A \in \mathcal{P}_j(B) \\ A \notin \mathcal{P}_j(B) \Longrightarrow A \notin \mathcal{P}_i(B) \end{gathered}$$

whenever $A, B \in \mathcal{A}_j \subset \mathcal{A}_i$.

One might hope for the existence of $(X, \mathcal{A}, \mathcal{P})$ with $\mathcal{A}$ the coarsest σ-algebra containing $(\mathcal{A}_i)_{i \in I}$ and $\mathcal{P}_i$ the restriction of $\mathcal{P}$ to $\mathcal{A}_i$, where the independence condition holds for $\mathcal{P}$, but just the observation that

$$\frac{\alpha_i(X)}{\beta_i(X)}$$

may be different for different $i \in I$ shows that this is impossible.

The natural generalizations of stochastic independence will not be developed here. (See conjecture 4 page 206).

Two other extensions of $(\mathcal{P}_i)_{i \in I_o}$ will however be introduced

Definition 153 ($\mu(I_0)$) *For a given $(\mathcal{P}_i)_{i \in I_o}$ we define*

$$\begin{gathered} \mu(I_0) = \{(A,B) \in \mathcal{A} \times \mathcal{A} \,|\, A \supset A_0, B \subset B_0, A_0 \in \mathcal{P}_i(B_0), \\ A_0, B_0 \in \mathcal{A}_i \textit{ for some } i \in I_0\} \\ \cup \\ \{(A,B) \in \mathcal{A} \times \mathcal{A} \,|\, A \subset A_0, B \supset B_0, A_0 \notin \mathcal{P}_i(B_0), \\ A_0, B_0 \in \mathcal{A}_i \textit{ for some } i \in I_0\} \end{gathered}$$

$\mu(I_0)$ is a somewhat strange system of subsets, it depends on the relations $(\mathcal{P}_i)_{i\in I_o}$, but it is the natural set on which to define the monotonic extension of $(\mathcal{P}_i)_{i\in I_o}$. It is obviously not an algebra, but it contains $\bigcup_{i\in I_0} \mathcal{A}_i \times \mathcal{A}_i$.

Definition 154 **($\sigma(I_0)$)** *The coarsest σ-algebra containing $(\mathcal{A}_i)_{i\in I_0}$ is denoted $\sigma(I_0), I_0 \subset I$.*

Notation 8 *A combined preference on $\mu(I_0)$ based on $(\mathcal{P}_i, \mathcal{A}_i)_{i\in I_0}$ will be denoted $\mathcal{P}_{mI_0}$*

Notation 9 *A combined preference on $\sigma(I_0)$ based on $(\mathcal{P}_i, \mathcal{A}_i)_{i\in I_0}$ will be denoted $\mathcal{P}_{cI_0}$*

First the monotonic extension can be defined first on $\bigcup_{i\in I_0} \mathcal{A}_i \times \mathcal{A}_i$ and then on all of $\mu(I_0)$.

Definition 155 (Monotonic extension) *$\mathcal{P}_{mI_0}$ is the monotonic extension of $(\mathcal{P}_i)_{i\in I_0}$ if it is the smallest monotone relation on $\mu(I_0)$ containing $\mathcal{P}_i$ for $i \in I_0$.*

Defining the combined relation $\mathcal{P}_{mI_0}$ on $\mu(I_0)$ as the monotonic extension of $(\mathcal{P}_i)_{i\in I_0}$ we have for

$$(A, B) \in \bigcup_{i\in I_0} \mathcal{A}_i \times \mathcal{A}_i$$

$$\begin{gathered} A \in \mathcal{P}_{I_0}(B) \Longleftarrow A \in \mathcal{P}_i(B) \Longleftrightarrow \lambda_i(A) > \lambda_i(B) \\ \text{for some } i \in I_0, \text{ for } A, B \in \mathcal{A}_i \text{ and all } \lambda_i \text{ with} \\ p_i \leqq \lambda_i \leqq b_i. \ \lambda_i : \mathcal{A}_i \to [0,1] \end{gathered}$$

and for

$$A, B \in \mu(I_0) \setminus \bigcup_{i\in I_0} \mathcal{A}_i \times \mathcal{A}_i$$

conditions given below in (14.1)

The monotonic extension $\mathcal{P}_m$ of in $(\mathcal{P}_i)_{i\in I_0}$ can be characterized by

$$\begin{gathered} A \in \mathcal{P}_m(B) \Leftrightarrow \text{if } \exists i \in I_0, A_0 \subset A, B_0 \supset B, \ A_0, B_0 \in \mathcal{A}_i \\ \text{such that } A_0 \in \mathcal{P}_i(B_0) \\ [A_0 \in \mathcal{P}_i(B_0) \Longleftrightarrow \lambda_i(A_0) > \lambda_i(B_0) \\ \text{for } \forall \lambda_i \text{ with } p_i \leqq \lambda_i \leqq b_i. \ \lambda_i : \mathcal{A}_i \to [0,1]] \end{gathered} \tag{14.1}$$

Definition 156 **($\Gamma(I_0)$)** *The class of measures defined on all of $\sigma(I_0)$ by*

$$\Gamma(I_0) = \{\lambda \,|\alpha_i(A) \leqq \lambda(A) \leqq \beta_i(A), A \in \mathcal{A}_i, i \in I_0\}$$
$$\Gamma(I_0) = \bigcap\nolimits_{i \in I_0} \Gamma(i), I_0 \subset I$$

$\Gamma(I_0)$ is not a very useful way of combining the knowledge in I_0. It depends on the arbitrary normalization of $(\alpha_i, \beta_i)_{i \in I_0}$. Even if $(\alpha_i, \beta_i)_{i \in I_0}$ is normalized for example by $\alpha_i(X) + \beta_i(X) = 2$, and this gives a well-defined $\Gamma(I_0)$, this set will have strange properties and presumably would not be the combination of knowledge wanted in applications.

The following class of measures will be used to express the combined knowledge in $(\alpha_i, \beta_i)_{i \in I_0}$. It is an assumption for this combination of knowledge that the combined knowledge can be expressed using a set of probability measures. Without some such extra assumptionn only the sets which can be compared directly for some $i \in I$ can be compared.

Definition 157 **($\Lambda(I_0)$)** *The class of probability measures defined on all of $\sigma(I_0)$ by*

$$\Lambda(I_0) = \{\lambda \,|p_i(A) \leqq \lambda(A) \leqq b_i(A), A \in \mathcal{A}_i, i \in I_0\}$$
$$\Lambda(I_0) = \bigcap\nolimits_{i \in I_0} \Lambda(i), I_0 \subset I$$

Remark 55 $\Gamma(I_0)$ *and* $\Lambda(I_0)$ *are a convex classes of measures, but convex sets with very special properties, and not any convex set of measures.*

Definition 158 **($\mathcal{P}_{cI_0}$)** *The extension $\mathcal{P}_{cI_0}$ of $(\mathcal{P}_i)_{i \in I_0}$ is defined on $\sigma(I_0)$ by*

$$A \in \mathcal{P}_{cI_0}(B) \Leftrightarrow \lambda(A) > \lambda(B), \forall \lambda \in \Lambda(I_0)$$

Definition 159 **($\mathcal{P}_c$)** *The extension $\mathcal{P}_c$ of $(\mathcal{P}_i)_{i \in I}$ is defined by*

$$\begin{array}{c} A \in \mathcal{P}_c(B) \Leftrightarrow \exists I_0 \\ \textit{such that } \lambda(A) > \lambda(B) \,\forall \lambda \in \Lambda(I_0) \end{array} \tag{14.2}$$

.

Theorem 60 *Assuming that a combined preference given $(\mathcal{P}_i)_{i \in I}$ and $(\mathcal{P}_i)_{i \in I_0}$ can be expressed by a set of measures, $\mathcal{P}_{cI_0}$ and $\mathcal{P}_c$ are* *the coarsest relations finer than the given individual relations.* $(\mathcal{P}_i)_{i\in I}$, $\mathcal{P}_{cI_0}$ *and* $\mathcal{P}_c$ *are monotone. Assuming consistency the relation on* $A, B \in \sigma(I_1)$ *is independent of* $\mathcal{P}_i$ *for* $\mathcal{A}_i \supset \sigma(I_1)$ *and* $\mathcal{P}_c = \mathcal{P}_{cI}$.

Proof. $\mathcal{P}_{cI_0}$ is by definition the coarsest relation expressed in terms of subsets of measures finer than $\mathcal{P}_i$ for all $i \in I_0$. Relations expressed in terms of intervals of measures are automatically monotone. Consistency means that a $\mathcal{P}_i$ on an $\mathcal{A}_i$ can not give new restrictions on an $\mathcal{A}_j$ with a relation $\mathcal{P}_j$ when $\mathcal{A}_j \subset \mathcal{A}_i$ ■

The independence assumption holds for the relations $(\mathcal{P}_i)_{i\in I}$. This has consequences for the properties of $\mathcal{P}_{cI_0}$ and $\mathcal{P}_c$. There may be sets A, B, A', B', C such that the independence condition page 98 is violated. The sets A, B, A', B' may be elements in a coarse $\sigma-$algebra with small or no uncertainty and the sets $A \cap C, A \setminus C$, etc. may only be elements in a finer $\sigma-$algebra with more uncertainty. $\mathcal{P}_{cI_0}$ can however not be such that it is represented by any convex set of measures. The fact that the individual constraints on the restrictions of the measures to sub $\sigma-$algebras are intervals gives special properties.

Given a $\mathcal{P}$ it should be possible to find conditions for the existence of $(\mathcal{A}_i)_{i\in I}$ and to construct $\Lambda(I)$, $\Gamma(I)$, and the relations $(\mathcal{P}_i)_{i\in I}$ and in that way find necessary and sufficient conditions for the representation (14.2).

But - to repeat - the independence assumption will not hold in general for $\mathcal{P}_{cI_0}$ and there does not exist α, β on $\mathcal{A}$ such that α_i, β_i can be recovered as restrictions of α, β to $\mathcal{A}_i$.

In order to compare two sets A and B given the knowledge $(\mathcal{P}_i)_{i\in I}$ without assuming that the knowledge can be combined to $\mathcal{P}_{cI_0}$ or to $\mathcal{P}_c$ the first step is to find an $i \in I$ such that A and B are $\mathcal{A}_i$-measurable. A is then preferred to B if $\lambda(A) > \lambda(B)$ for all $\lambda \in \Lambda(i)$.

The second step - assuming monotonicity - is to look for sets A_0 and B_0 contained in and containing A and B respectively such that A_0 is preferred to B_0 with the larger knowledge contained in the coarser σ-algebra containing A_0 and B_0.

If we assume that the knowledge $(\mathcal{P}_i)_{i\in I_0}$ can be combined to $\mathcal{P}_{cI_0}$ the situation is much simpler. A is then preferred to B if $\lambda(A) > \lambda(B)$ for all $\lambda \in \Lambda(I_0)$, where $\Lambda(I_0)$ expresses the combined knowledge in I_0. If all the knowledge is combined A is preferred to B if $\lambda(A) > \lambda(B)$ for all $\lambda \in \Lambda(I)$

If all the relations were total and transitive so (so $\alpha_i = \beta_i, i \in I$) the situation is of course even simpler. In that situation α_i can be determined from any $\lambda \in \Lambda$ just as the restriction of λ to $\mathcal{A}_i$. (See Strassen (1965) [171]).

Example 13 *1. To illustrate that the extension $\mathcal{P}_c$ gives more than the individual preferences we shall return to a slightly modified version of example 12. Let $X_1 = X_2 = \{0,1\}$ so the sample space is*

$$X_1 \times X_2 = \{(0,0),(0,1),(1,0),(1,1)\}$$

$$\mathcal{A}_1 = 2^{X_1}, \mathcal{A}_2 = 2^{X_2} \text{ and } \mathcal{A} = 2^{X_1 \times X_2}$$

Assume that the preferences on $\mathcal{A}_1$ and $\mathcal{A}_2$ are total orders and can be represented by probabilities $p_1, 1-p_1, p_2$, and $1-p_2$, so there is no uncertainty on X_1 or X_2. We may anyway want uncertainty on $\mathcal{A}$, the joint distribution, so the relation on $\mathcal{A}$ can be represented by

$$(\alpha_{ij}, \beta_{ij}), \alpha_{ij} \leqq \beta_{ij}, i,j \in \{0,1\}$$

or by

$$\begin{aligned} p_{ij} &= \frac{\alpha_{ij}}{\alpha_{ij}+\sum_{\nu \neq i,j} \beta_\nu} \\ b_{ij} &= \frac{\beta_{ij}}{\beta_{ij}+\sum_{\nu \neq i,j} \alpha_\nu} \end{aligned}$$

and for example

$$\begin{aligned} p(\{(0,0),(1,1)\}) &= \frac{\alpha_{00}+\alpha_{11}}{\alpha_{00}+\alpha_{11}+\beta_{00}+\beta_{11}} \\ b(\{(0,0),(1,1)\}) &= \frac{\beta_{00}+\beta_{11}}{\alpha_{00}+\alpha_{11}+\beta_{00}+\beta_{11}} \end{aligned}$$

and in general for S any subset of $X_1 \times X_2$

$$\begin{aligned} p(S) &= \frac{\alpha(S)}{\alpha(S)+\beta(S^c)} \\ b(S) &= \frac{\beta(S)}{\beta(S)+\alpha(S^c)} \end{aligned}$$

where $b_{ij} \geqq p_{ij} \geqq 0\,(i,j=1,2)$. (So $(1,1) \in \mathcal{P}_0(2,2) \Leftrightarrow p_{11} > b_{22}$ etc.).

Let now λ_{ij} be a measure for which all three sets of conditions hold so

$$\begin{array}{rcl} b_{ij} \geqq \lambda_{ij} \geqq p_{ij} & & \\ b(S) \geqq \lambda(S) \geqq p(S) & \Longleftrightarrow & \lambda \in \Lambda(0) \\ \lambda_{00} + \lambda_{01} = p_1 & \Longleftrightarrow & \lambda \in \Lambda(1) \\ \lambda_{00} + \lambda_{10} = p_2 & \Longleftrightarrow & \lambda \in \Lambda(2) \\ \lambda_{11} + \lambda_{01} = 1 - p_2 & \Longleftrightarrow & \lambda \in \Lambda(2) \\ \lambda_{10} + \lambda_{11} = 1 - p_1 & \Longleftrightarrow & \lambda \in \Lambda(1) \end{array}$$

consistency means

$$\begin{array}{ccccc} b(\{(0,0),(0,1)\}) & \geqq & p_1 & \geqq & p(\{(0,0),(0,1)\}) \\ b(\{(0,0),(1,0)\}) & \geqq & p_2 & \geqq & p(\{(0,0),(1,0)\}) \end{array}$$

Then $\lambda \in \Lambda(1,2) = \Lambda(1) \cap \Lambda(2)$ easily implies

$$\begin{array}{rlll} \lambda_{10} - \lambda_{01} = p_2 - p_1 & so & (1,0) \in \mathcal{P}_c(0,1) \Longleftrightarrow & p_2 - p_1 > 0 \\ \lambda_{00} - \lambda_{11} = p_1 + p_2 - 1 & so & (0,0) \in \mathcal{P}_c(1,1) \Longleftrightarrow & p_1 + p_2 > 1 \end{array}$$

and

$$\begin{array}{rl} (0,1) \in \mathcal{P}_c(1,0) \Longleftrightarrow & p_1 - p_2 > 0 \\ (1,1) \in \mathcal{P}_c(0,0) \Longleftrightarrow & p_1 + p_2 < 1 \end{array}$$

so unless $p_1 = p_2 = \frac{1}{2}$ some elements which could not be compared based on the three individual preferences can now be compared based on the combined preference relation $\mathcal{P}_c$. $(0,0)$ and $(0,1)$ or $(1,0)$ and $(1,1)$ can still only be compared based on $\mathcal{P}_0$.

For the case $p_1 + p_2 < 1$ and $p_2 - p_1 > 0$ we obtain

$$\begin{array}{c} \Lambda(1,2) = \left\{\lambda \in \mathbb{R}^4_+ \,|\lambda = t\,\lambda^1 + (1-t)\,\lambda^2\right\} \\ for\ t \in [0,1] \\ \lambda^1 = (0, p_1, p_2, 1 - (p_1 + p_2)) \\ \lambda^2 = (p_1, 0, p_2 - p_1, 1 - p_2) \end{array}$$

Other cases will give the non negative part of the straight line through the points $\lambda^1 \in \mathbb{R}^4$ and $\lambda^2 \in \mathbb{R}^4$

$$\Lambda(0) = \left\{\lambda \in \mathbb{R}^4_+ \,|p \leqq \lambda \leqq b\right\}$$

The complete characterization of $\Lambda(I)$ is

$$\Lambda(I) = \left\{\lambda \in \mathbb{R}^4_+ \,|\lambda = t\,\lambda^1 + (1-t)\,\lambda^2, p \leqq \lambda \leqq b\right\}$$

2. It will be instructive also to find the conditional measures in $\Lambda(I)$ given the first observation. For $x_1 = 0$ we get

$$\begin{gathered}(\Lambda(1,2)\,|x_1 = 0) = \\ \left\{\lambda \in \mathbb{R}_+^2 \,|\lambda = t\,(0,1) + (1-t)\,(1,0)\right\}\end{gathered}$$

only $\Lambda(0)$ gives constraints on the set of conditional measures. For $x_1 = 0$ we get

$$\begin{gathered}(\Lambda(1,2)\,|x_1 = 1) = \\ \left\{\lambda \in \mathbb{R}_+^2 \,\middle|\lambda = \tfrac{1}{1-p_1} t\,((p_2, 1-p_1-p_2) + (1-t)\,(p_2-p_1, 1-p_2))\right\}\end{gathered}$$

so $\mathcal{P}_2$ alone has more knowledge about the second observation x_2 than the knowledge after observing $x_1 = 0$ or $x_1 = 1$.

Remark 56 *The example shows that observations may increase the uncertainty. This may seem paradoxical. What the example shows is that taking conditional measures and normalizing does not commute. Conditional measures taken before combining gives one result and combining first gives another. If the first is done in this example only $\mathcal{P}_2$ would matter. Both approaches are mathematically possible and it is up to the applications to choose. Apparently it gives the nicest theory to accept the paradox i.e. to combine all the knowledge in the form of normalized measures and then take conditional measures. It least this is what will be done in chapter 18 in the suggested foundation of statistics.*

14.2.3 Y and Z general

The case $X = X_1 \times X_2$, with $(X, \mathcal{A}, \mathcal{G} \times \mathcal{H}, \mathcal{P})$, $(X_1, \mathcal{A}_1, \mathcal{G}_1 \times \mathcal{H}_1, \mathcal{P}_1)$ and $(X_2, \mathcal{A}_2, \mathcal{G}_2 \times \mathcal{H}_2, \mathcal{P}_2)$ will be considered first.

The $\mathcal{P}$ on $\mathcal{G} \times \mathcal{H}$ obviously determines relations on $\mathcal{G}_1 \times \mathcal{H}_1$ and on $\mathcal{G}_2 \times \mathcal{H}_2$ (assuming that functions measurable with respect to the sub σ-algebras $\mathcal{A}_1$ and $\mathcal{A}_2$ of $\mathcal{A}$ are in $\mathcal{G} \times \mathcal{H}$), but from the point of view of applications there is no need to assume that these relations are $\mathcal{P}_1$ and $\mathcal{P}_2$. $\mathcal{P}$'s restriction to $\mathcal{G}_1 \times \mathcal{H}_1$ and $\mathcal{G}_2 \times \mathcal{H}_2$ would usually be coarser than $\mathcal{P}_1$ and $\mathcal{P}_2$.

The representations will for the case of state independence and integral representation be

$$f(g,h,A) = \int_A W(g(x), h(x))\, d\gamma$$
where
$$f : \mathcal{G} \times \mathcal{H} \times \mathcal{A} \to \mathbb{R}$$
$$W : Y \times Z \to \mathbb{R},\ \gamma : \mathcal{A} \to \mathbb{R}$$
and
$$f_1(g_1, h_1, A_1) = \int_{A_1} W_1(g_1(x_1), h_1(x_1))\, d\gamma_1$$
where
$$f_1 : \mathcal{G}_1 \times \mathcal{H}_1 \times \mathcal{A}_1 \to \mathbb{R}$$
$$W_1 : Y \times Z \to \mathbb{R},\ \gamma_1 : \mathcal{A}_1 \to \mathbb{R}$$
and
$$f_2(g_2, h_2, A_2) = \int_{A_2} W_2(g_2(x_2), h_2(x_2))\, d\gamma_2$$
where
$$f_2 : \mathcal{G}_2 \times \mathcal{H}_2 \times \mathcal{A}_2 \to \mathbb{R}$$
$$W_2 : Y \times Z \to \mathbb{R},\ \gamma_2 : \mathcal{A} \to \mathbb{R}$$

Natural conditions will imply (see chapter 12 page 169) that the preference functions can be chosen identical i.e. $W_2 = W_1 = W$.

The general case $(X, \mathcal{A}, \mathcal{G} \times \mathcal{H}, \mathcal{P})$, $(X_i, \mathcal{A}_i, \mathcal{G}_i \times \mathcal{H}_i, \mathcal{P}_i)_{i \in I}$ gives rise to the representation of a $\mathcal{P}$ on functions defined on $(X, \sigma(I))$ with values in $Y \times Z$

$$f : \mathcal{G} \times \mathcal{H} \to \mathbb{R}$$

represents $\mathcal{P}$ i.e.

$$f(g,h) > 0 \Leftrightarrow g \in \mathcal{P}(h)$$
where
$$f(g,h) = \inf_{\lambda \in \Lambda(I)} \int w(g_0(x), h_0(x))\, d\lambda$$
or
$$\int w(g_0(x), h_0(x))\, d\lambda > 0 \text{ for } \lambda \in \Lambda(I)$$

Conditions for this representation are consistency, state independence, and independence of the $\mathcal{P}_i$.

Remark 57 *Similar results can be obtained without the state independence assumption.*

This of course is more complicated than just taking the expected value of two utility functions and compare, but given that uncertainty is needed in applications there seems to be no way to avoid the complications.

14.3 Two level uncertainty

Integrals of a real function of functions of two variables can be viewed as a representation of a total preorder on a space of functions of two variables. Fubini's theorem[2] tells us that there for each fixed value of one of the variables is a total preorder on functions of the other variable, that there are measures on each of the factors representing the knowledge on this factor, and that the three possible ways of getting the expected utility of a function of two variables all give the same result. Generalizing this result to uncertainty is very important for any theory hoping to generalize the Savage theory.

This way of looking at Fubini's theorem can be expressed as follows. Let $\mathcal{G} \subset Y^{X_1 \times X_2}$, $\mathcal{G}(\cdot, x_2) \subset Y^{X_1}$ and $\mathcal{G}(x_1, \cdot) \subset Y^{X_2}$ be sets of measurable functions

Space	Representation	Given
$(\mathcal{G}, \succsim)$	$\int u(g(x_1, x_2), x_1, x_2)\, d\lambda$	
$[\mathcal{G}(x_1, \cdot), \succsim_{x_1}]$	$\left[\int u_2(g(x_1, x_2), x_1, x_2)\, d\lambda_2\right]$	x_1
$\langle \mathcal{G}(\cdot, x_2), \succsim_{x_2} \rangle$	$\left\langle \int u_1(g(x_1, x_2), x_1, x_2)\, d\lambda_1 \right\rangle$	x_2

(14.2)

$(\mathcal{G}, \succsim)$ is a totally preordered space of functions defined on $X_1 \times X_2$ with values in Y. If the conditions for theorems 26, 27 (page 95), and 52 (page 155), then the relation can be represented by the expected value of a utility u function on $Y \times X_1 \times X_2$ with respect to a measure λ on $X_1 \times X_2$.The representation is determined up to multiplication with a positive constant and addition of a constant. u and λ is for a fixed representation determined up to a Radon Nikodým derivative.

$[\mathcal{G}(x_1, \cdot), \succsim_{x_1}]$ is for each x_1 a totally preordered space of functions defined on X_2 with values in Y.

$\langle \mathcal{G}(\cdot, x_2), \succsim_{x_2} \rangle$ is for each x_2 a totally preordered space of functions defined on X_1 with values in Y. Under the same conditions these relation can be represented as in 14.2. u_1 and λ_1 (and u_2 and λ_2) are as before determined up to a Radon Nikodým derivative.

---

[2] See for example Loève (1960) [126] page 136 or Halmos (1961) [93] page 148.

In addition there is a total preorder on the set of functions defined on X_1 with values in the space

$$[\mathcal{G}(x_1,\cdot), \succsim_{x_1}] \text{ i.e. } \left([\mathcal{G}(x_1,\cdot), \succsim_{x_1}]^{X_1}, \succsim_1\right)$$

and a preorder on the set of functions defined on X_2 with values in the space

$$\langle \mathcal{G}(\cdot, x_2), \succsim_{x_2}\rangle \text{ i.e. } \left(\langle \mathcal{G}(\cdot, x_2), \succsim_{x_2}\rangle^{X_2}, \succsim_2\right)$$

These spaces are mean groupoids and the consistency conditions for a representation for these total preorders are known from theorem 8.10 page 103. The mapping from equivalence classes in

$$[\mathcal{G}(x_1,\cdot), \succsim_{x_1}]$$

to equivalence classes in

$$\left([\mathcal{G}(x_1,\cdot), \succsim_{x_1}]^{X_1}, \succsim_1\right)$$

should be consistent in the sense that it is a homomorphism not only in the order but also in mean groupoid structure which under the independence condition can be defined on the spaces.

Fubini's theorem then tells us that

$(\mathcal{G}, \succsim) = \left([\mathcal{G}(x_1,\cdot), \succsim_{x_1}]^{X_1}, \succsim_1\right) = \left(\langle \mathcal{G}(\cdot, x_2), \succsim_{x_2}\rangle^{X_2}, \succsim_2\right)$
$\int u(g(x_1, x_2), x_1, x_2)\, d\lambda =$
$\int \left[\int u(g(x_1, x_2), x_1, x_2)\, d\lambda_2\right] d\lambda_1 =$
$\int \left\langle \int u(g(x_1, x_2), x_1, x_2)\, d\lambda_1 \right\rangle d\lambda_2$

(Fubini)

Expressed in this way Fubini's theorem shows that (u_1, λ_1) and (u_2, λ_2) can be chosen such that $u_1 = u_2 = u$ and that the three approaches are equivalent. The knowledge-preference on a space of functions can thus be expressed in three different ways either directly on the space of functions without using the product structure of the domain of the functions, or - in two different ways - as a conditional preferences on functions defined on one of the factors and then in addition knowledge, preference on the other factor.

One of the expressions is enough to give the representation and the other two will not add anything.

This equivalence does not generalize to non-total relations. Replacing the total preorders by non-total relations but keeping the other assumptions will as shown earlier in this book result in the same type of representations, but the Fubini result will have to be generalized.

Let $\mathcal{G} \subset Y^{X_1 \times X_2}, \mathcal{G}(\cdot, x_2) \subset Y^{X_1}$ and $\mathcal{G}(x_1, \cdot) \subset Y^{X_2}$ be sets of measurable functions. With relations $\mathcal{P}, \mathcal{P}_{x_1}, \mathcal{P}_{x_2}, \mathcal{P}_1$ and $\mathcal{P}_2$ as follows

$$(\mathcal{G}, \mathcal{P}) ; \left(\left[\mathcal{G}(x_1, \cdot), \mathcal{P}_{x_1} \right]^{X_1}, \mathcal{P}_1 \right) ; \left(\left\langle \mathcal{G}(\cdot, x_2), \mathcal{P}_{x_2} \right\rangle^{X_2}, \mathcal{P}_2 \right)$$

Space	Representation
$(\mathcal{G}, \mathcal{P})$	
$[\mathcal{G}(x_1, \cdot), \mathcal{P}_{x_1}]$	$f_{x_1}(g, h, X_2) =$ $\left[\int W_2(g(x_1, x_2), h(x_1, x_2), x_1, x_2)\, d\gamma_2\right]$
$\langle \mathcal{G}(\cdot, x_2), \mathcal{P}_{x_2} \rangle$	$f_{x_2}(g, h, X_1) =$ $\left\langle \int W_1(g(x_1, x_2), h(x_1, x_2), x_1, x_2)\, d\gamma_1 \right\rangle$
$\left([\cdots]^{X_1}, \mathcal{P}_1 \right)$	See equation 14.3 page 218
$\left(\langle \cdots \rangle^{X_2}, \mathcal{P}_2 \right)$	

$(\mathcal{G}, \mathcal{P})$ is a space of functions defined on $X_1 \times X_2$ with values in Y.

$[\mathcal{G}(x_1, \cdot), \mathcal{P}_{x_1}]$ is for each x_1 a space of functions defined on X_2 with values in Y. The relation $\mathcal{P}_{x_1}$ can be represented by the expected value of a preference function on $Y \times \{x_1\} \times X_2$ with respect to a measure on $\{x_1\} \times \mathcal{X}_2$.

$\langle \mathcal{G}(\cdot, x_2), \mathcal{P}_{x_2} \rangle$ is for each x_2 a space of functions defined on X_1 with values in Y. The relation $\mathcal{P}_{x_2}$ can used to define a commutative mean groupoid with zero and can be represented by the expected value of a preference function on $Y \times Y \times X_1 \times \{x_2\}$ with respect to a measure on $\mathcal{X}_1 \times \{x_2\}$.

The conditions for these two representations are the conditions from theorem 10 page 115 in the special case of the integral representation equation 52 page 155.

In addition there are relations on a set of functions defined on X_1 with values in the space $[\mathcal{G}(x_1, \cdot), \mathcal{P}_{x_1}]$ i.e. $\left([\mathcal{G}(x_1, \cdot), \mathcal{P}_{x_1}]^{X_1}, \mathcal{P}_1 \right)$

and on the set of functions defined on X_2 with values in the space $\langle \mathcal{G}(\cdot, x_2), \mathcal{P}_{x_2} \rangle$ i.e. $\left(\langle \mathcal{G}(\cdot, x_2), \mathcal{P}_{x_2} \rangle^{X_2}, \mathcal{P}_2 \right)$. The first of these relations can for the case where $\mathcal{P}_1$ is a consistent relation be represented by

$$\begin{array}{c} \int f_2 \left(g\left(x_1, \cdot\right), h\left(x_1, \cdot\right), x_1, X_2 \right)^+ d\alpha_1 - \\ \int f_2 \left(g\left(x_1, \cdot\right), h\left(x_1, \cdot\right), x_1, X_2 \right)^- d\beta_1 \end{array} \tag{14.3}$$

The conditions for this representation are known from subsection 9.7.5 page 126 in particular theorem 44 page 126 and the representations ✠ and ✠✠. The mapping from equivalence classes in

$$[\mathcal{G}(x_1, \cdot) \times \mathcal{G}(x_1, \cdot), gr\mathcal{P}_{x_1}]$$

to equivalence classes in

$$\left([\mathcal{G}(x_1, \cdot) \times \mathcal{G}(x_1, \cdot), gr\mathcal{P}_{x_1}]^{X_1}, \mathcal{P}_1 \right)$$

should be consistent in the sense that it is a homomorphism not only in the order but also in the mean groupoid structure which under the independence condition can be defined on the spaces of equivalence classes. (See chapters 8 and 9).

It is of course important to know under which conditions on a $\mathcal{P}$ we can get a Fubini like equivalence between the formulation in terms of $\mathcal{P}$ and a formulation in terms of $(\mathcal{P}_{x_1}, \mathcal{P}_1)$. This requires a decomposition theorem copied after the Hahn decomposition theorem and an independence conditions on $\mathcal{P}$ which will be called copositive independence.

Definition 160 (semiindependent) *$\mathcal{P}$ is semi-independent with respect to $\mathcal{X}_1$ if for $A \in \mathcal{X}_1$*

$$\begin{array}{c} \mathcal{P}_{Ah} : \mathcal{G}\,|A \to 2^{\mathcal{G}|A} \text{ defined by} \\ g\,|A \in \mathcal{P}_{Ah}\left(g'\,|A\right) \text{ if } \left(g \boxtimes_A h\right) \in \mathcal{P}\left(g' \boxtimes_A h\right) \end{array}$$

is independent of h

Theorem 61 (decomposition) *Let $\mathcal{P}$ be semi-independent with respect to $\mathcal{X}_1$and continuous with respect to a measure μ on $\mathcal{X}_1$. Then there exists for any pair $g, g' \in \mathcal{G}$ a partition (B, B') of X_1 $(B \in \mathcal{X}_1)$ such that*

$$\begin{array}{c} A \subset B \Rightarrow g \in \mathcal{P}_A\left(g'\right) \\ A \subset B' \Rightarrow g \notin \mathcal{P}_A\left(g'\right) \end{array}$$

Proof. Proof of the Hahn decomposition can be found in many texts. See Halmos (1950) [93] page 121, or Loève (1960) [126] page 86. With small modifications the same proof can be applied here. Define for example in the proof of the Hahn decomposition theorem in Dunford Schwartz (1958) ([64] page 129-130) the set P by

$$\{E \in \mathcal{X}_1 \,|(g\,|A \cap E) \in \mathcal{P}\,(h\,|A \cap E)\,, \text{ for all } A \in \mathcal{X}_1\}$$

Then the rest of the proof is almost the same as in [64]. ∎

Remark 58 *For a given pair (g, g') the decomposition is unique up to a $\mu-$null set*

Definition 161 (copositive pairs) *Two pairs of functions (g, g') and (h, h') are* ***copositive*** *with respect to $\mathcal{P}$ if the "Hahn" decomposition of the two pairs are essentially identical.*

Definition 162 (copositive independence) *$\mathcal{P}$ is copositive independent if the independence condition 114 page 98 holds for graph$\mathcal{P}$ on copositive pairs of functions.*

Theorem 62 (Two level uncertainty) *Let $\mathcal{P}$ be semi-indepen dent with respect to $\mathcal{X}_1$ and continuous with respect to a measure μ on $\mathcal{X}_1$.*

$$(\mathcal{G}, \mathcal{P})$$

with copositive independence is equivalent to

$$\left(\left[\mathcal{G}\,(x_1, \cdot\,)\,, \mathcal{P}_{x_1}\right]^{X_1}, \mathcal{P}_1\right)$$

where $(\mathcal{P}_{x_1})_{x_1 \in X_1}$ and $\mathcal{P}$ are consistent, $\mathcal{P}_1$ is a consistent relation, and the independence conditions holds for $(\mathcal{P}_{x_1})_{x_1 \in X_1}$ and $\mathcal{P}_1$. Both can be represented by

$$\begin{array}{l} \int f_2\,(g\,(x_1, \cdot\,)\,, h\,(x_1, \cdot\,)\,, x_1, X_2)^+ \,d\alpha_1 \\ -\int f_2\,(g\,(x_1, \cdot\,)\,, h\,(x_1, \cdot\,)\,, x_1, X_2)^- \,d\beta_1 \end{array} \qquad \text{(2 level. unc.)}$$

$\left(\left[\mathcal{G}\,(x_1, \cdot\,)\,, \mathcal{P}_{x_1}\right]^{X_1}, \mathcal{P}_1\right)$ determines $(\mathcal{G}, \mathcal{P})$ but not conversely.

Proof. Let $B \in \mathcal{X}_1$ be any non null subset, where also $B' = B^c$ is non null (both $\mathcal{P}$ non null sets). Define

$$(\mathcal{G} \times \mathcal{G})_B = $$
$$\{(g, h) \in \mathcal{G} \times \mathcal{G}\,|(B, B') \text{ is a Hahn decomposition for } (g, h)\ \}$$

The independence condition will now hold on

$$(\mathcal{G} \times \mathcal{G})_B \,|B \times (\mathcal{G} \times \mathcal{G})_{B'} \,|B'$$

The representations (f_2, α_1) on $((\mathcal{G} \times \mathcal{G})_B \,|B)$ and on $\mathcal{A}\,|B$ and (f_2, β_1) on $(\mathcal{G} \times \mathcal{G})_{B'} \,|B'$ and on $\mathcal{A}\,|B'$ can now be obtained from theorems 26 and 52. Given the choice of (f_2, α_1, β_1) for one B there is obviously a unique extension (f_2, α_1, β_1) to all of $\mathcal{G} \times \mathcal{G}$ and $\mathcal{A}$ ■

(f_2, α_1, β_1) are only determined up to a Radon-Nikodým derivative (see remark 35 page 105) (and of course a positive constant). α_1 and β_1 can even be determined separately, and this is the reason that $(\mathcal{G}, \mathcal{P})$ does not determine $\left([\mathcal{G}(x_1, \cdot), \mathcal{P}_{x_1}]^{X_1}, \mathcal{P}_1\right)$. Given $(\mathcal{G}, \mathcal{P})$ many choices of $\left([\mathcal{G}(x_1, \cdot), \mathcal{P}_{x_1}]^{X_1}, \mathcal{P}_1\right)$ will result in the same relation on $\mathcal{G}$, the different choices will result in the same conditional preference $\mathcal{P}_{x_1}$ for each x_1, but will have different preorders $\mathcal{P}_1$ and different α_1, β_1, f_2. The natural limits on the choice are $\alpha_1 \leqq \beta_1$ and $f_2(y', y'', x_1, x_2) + f_2(y'', y', x_1, x_2) \leqq 0$ for all $(x_1, x_2) \in X_1 \times X_2$, this means that the uncertainty both on Y and on $\mathcal{A}$ is non-negative.

14.4 Conclusions

The conclusion is that there are unavoidable complications in combining uncertainty compared to combining probabilities, but that there are solutions to the problems. The mathematics used to get representations is the same, and the representations are obvious generalizations of the well-known representations, but the reduction of all knowledge to one probability measure, which can be used to find utilities as expected values no longer holds. It is necessary to use all the knowledge to give inequalities and it is necessary to specify both preferences and conditional preferences. On the other hand the only set functions showing up in the representations are measures i.e. additive.

14.5 Note

The results in this chapter are new.

15

Conditional uncertainty

15.1 Introduction

The fact that all representation theorems in this book contain measures and not more general non-additive set functions has the extremely convenient consequence that conditional relations have very natural representations via the conditional measures. Combining relations as in chapter 14 and taking conditional measures does however not commute. It is of course mathematically possible to perform the operations in any order and the results will be different. See example 13 and remark 56 page 211.

15.2 Relations on function spaces

Theorem 63 (theorem 36) *Let $\mathcal{G}$ be a mixture with respect to an algebra $\mathcal{A}$ where $\mathcal{A}$ has more than three disjoint non-null sets. Let Q be independent, open and essential for all partitions*

$$(A_i)_{i\in I}, (A_i \in \mathcal{A})$$

of X and assume $\mathcal{G}\,|A_i$ connected. Then there exists a function

$$\begin{gathered} f : \mathcal{G} \times \mathcal{A} \to \mathbb{R} \\ \text{such that} \\ f(g, \cdot) \text{ is additive} \\ \text{and} \\ f(g, X) > 0 \Leftrightarrow g \in Q \end{gathered}$$

and all the results in chapters 9, 11, 12, and 13 which are special cases of that theorem, can by the Radon-Nykodým theorem be supplemented by

Theorem 64 (conditional representation) *Let $\mathcal{A}$ be a $\sigma-$algebra, let $\mathcal{B}$ be a sub $\sigma-$algebra, and let f be any $\sigma-$additive function from the theorems in chapters 9, 11, 12, and 13 - for example*

$$f : \mathcal{G} \times \mathcal{A} \to \mathbb{R}$$

with $\gamma(g,\cdot) = f(g,\cdot)$ on $\mathcal{B}$, then there exists a $\mathcal{B}-$ a.e uniquely determined function

$$\widetilde{f} : \mathcal{G} \times \mathcal{A} \times X \to \mathbb{R}$$

such that $\widetilde{f}(g,A,\cdot)$ is $\mathcal{B}-$measurable and

$$f(g,A) = \int_A \widetilde{f}(g,A,x)\, d\gamma(g)$$

Proof. Direct application of the Radon-Nikodým theorem ■

Remark 59 *With the assumptions from chapter 11 the γ can be chosen independent of g.*

Conditional representations for the other representations follow in an analogous way, and will not be covered explicitly.

15.3 One probability-uncertainty measure

The special case of an uncertainty space will be considered explicitly.

Let $(X,\mathcal{A},\mathcal{P})$ be an uncertainty space represented by (α,β) (or (λ,μ) or (p,b)) and let $\mathcal{B} \subset \mathcal{A}$. If $B \in \mathcal{B}$ and $\lambda(B) > 0$, then $\alpha(A,B) = \alpha(A \cap B)$ and $\beta(A,B) = \beta(A \cap B)$ gives the obvious uncertainty on $\mathcal{A}$ given B. One gets in general the existence of

$$\begin{gathered}\alpha : \mathcal{A} \times X \to \mathbb{R},\ \beta : \mathcal{A} \times X \to \mathbb{R}\\ \text{with } \alpha(A,\cdot),\ \beta(A,\cdot)\ \mathcal{B}-\text{measurable}\\ \text{and } \alpha(A) = \int \alpha(A,x)\, d\gamma,\ \text{and } \beta(A) = \int \beta(A,x)\, d\gamma\end{gathered}$$

The choice of γ is arbitrary, any other measure on $\mathcal{B}$ with respect to which (α,β) is absolutely continuous would work (i.e. give the same relation $\mathcal{P}(\cdot,x)$ on $\mathcal{A}$ given x).

A natural normalization is $\alpha(A,\cdot) + \beta(A,\cdot) = 2$, so the conditional probability measure is normalized to 1.

15.4 Several probability-uncertainty measures

15.4.1 $Y = Z = \{0, 1\}$

In this case the measures involved in the representation can just be replaced by the conditional measures and the representation repeated. As is shown in example 13 the conditional measures can be taken before combining or after. For one-level uncertainty problems the most satisfactory theory is probably obtained by combining the knowledge expressed in the individual pairs of measures on sub $\sigma-$ algebras first and then take the conditional measures. Some applications and the two-level uncertainty may want the reverse order.

15.4.2 Y and Z general

In this case the measures involved in the representation can just be replaced by the conditional measures and the representation repeated.

15.5 Two level uncertainty

Here the situation is more complicated. See chapter 18 theorem 70 page 245.

15.6 Conclusion

This chapter gives the mathematics of the conditional representations. It is obviously a question for applications if the conditional representations also are the representations of for example the preferences given an observation.

Part VI

Applications

16

Production, utility, preference

16.1 Introduction

The results in this book have of course many applications. Whenever an integral or a sum is used, the use may be justified - via the independence assumption - as a representation of a subset, a total preorder, or just a relation.

16.2 Production functions

Let n be a set of commodities - inputs and outputs - involved in the production of a firm and let

$$Y \subset \prod_{j \in N} S_j \subset \mathbb{R}^n, \text{ where } S_j \subset \mathbb{R}^{n_j}, \sum n_j = n$$

be the production sets for the firm (i.e. the set of points in the commodity space which can be produced by the firm). Corollary 13 page 117 applied to Y^c will then yield functions $f_j : \mathbb{R}^{n_j} \to \mathbb{R}$ such that

$$y \in Y \Leftrightarrow \sum f_j(y_j) \geqq 0$$

The only problem is therefore to give an economic interpretation of the assumptions.

The independence assumption for Y with respect to $A \subset N$ can be written

$$\begin{array}{ll} (y_A, y_{A^c}) \in Y & (y'_A, y_{A^c}) \notin Y \\ (y_A, y'_{A^c}) \notin Y & (y'_A, y'_{A^c}) \in Y \end{array} \qquad \text{(NOT)}$$

can not hold for any choice of $(y_A, y_{A^c}) \in Y$ and $(y'_A, y'_{A^c}) \in Y$.

It means that if a change in the production within the commodities in A from y'_A to y_A for some value y_{A^c} of the production of commodities outside A result in a change from a point outside to a point inside the production set, then the same change inside A can not for another value y'_{A^c} result in the opposite change. If y_A is output and y_{A^c} input $(y_A \geqq 0, y_{A^c} \leqq 0)$, then the independence assumption implies the set of possible outputs for two different inputs are either identical or one is contained the other.

The interpretation of essential is obvious. Y closed in the usual topology on $\mathbb{R}$ and $(S_j)_{j \in N}$ connected will imply Y^c open. So to summarize

Theorem 65 *Let $\#N > 3$, let Y be a closed, essential, and independent subset of $\prod_{j \in N} S_j$,where S_j are connected subset of $\mathbb{R}^{N_j}$ $(j \in N)$.Then there exist continuous functions $f_j : S_j \to \mathbb{R}$ such that*

$$y \in Y \Leftrightarrow \sum f_j(y_j) \geqq 0$$

Proof. Use corollary 13 on Y^c and set $f_j = -u_j$ ■

Remark 60 *The assumption that $S_j \subset \mathbb{R}^{n_j}$ is only made to have a convenient way of expressing the topological assumptions. It is not needed for the theorem. The factors of the production set can be arbitrary sets.*

16.3 Additive preference functions

The earlier representation and decomposition theorems for preferences on function spaces and on finite products have the obvious interpretations representations and pure preference and uncertainty on commodity spaces. The independence assumption has also a clear interpretation. The special properties of demand with additive preference functions will not be developed here.

16.4 Additive utility functions

Also the special case of total preorders and therefore utility function as representation are just immediate interpretations of the earlier results.

16.5 Notes

The *Handbook of Utility Theory* (1998) (edited by Barbera, Hammond and Seidl) [16] is a good reference for utility theory. Chapter 2 (by Blackorby, Primont and Russell) in [16] discusses additive utility functions.

17

Preferences over time

17.1 Introduction

One of the most unsatisfactory assumptions in economic theory in general and in general equilibrium theory in particular is the assumption that consumers have total preorders as preferences for consumption in future periods. The results in this book allows for representation of preferences that are not total preorders. The direct interpretation of the results in this book gives the first set of results about preferences over time.

The fact that time is a set with a structure and an interpretation of its own and not just an arbitrary set with a $\sigma-$algebra, but a set with an order relation and a metric, means that three different questions arise.

1. Does the special structure of time allow for special assumptions with special cases of the representations theorems as a result? In particular how do results about stationarity generalize?

2. Is the independence condition for the general representations theorems acceptable. In particular is "uncertainty about the time preference" at one point in time independent of the uncertainty in neighboring points or of the uncertainty for later points in time?

3. Is the automatic updating of preferences as conditional preferences what we want for example as preferences at a later point in time?

Given that these three classes of problems have found satisfactory solutions new questions arises. What are the consequences for the existing theory based on total preorders as preferences? Can optimal control theory still be used to give conditions for optimality? How does general equilibrium theory change, if the preferences at time t are not the conditional preferences from time 0? Can in the same economy two different growth rates result because of the uncertainty?

This chapter will concentrate on presenting the direct application of the main representation theorem for uncertainty, and then to present and discuss what is a more general and possibly a more applicable result.

17.2 $((T,\mathcal{A}),Y,Z,\mathcal{G},\mathcal{H},\mathcal{P})$ Existence of $f:\mathcal{G}\times\mathcal{H}\times\mathcal{A}\rightarrow\mathbb{R}$

Let $T=\mathbb{R}_+$ or $T=\{0,1,2,\cdots,n\}$ (in the interpretation time), Y and Z are arbitrary sets and $\mathcal{A}$ a system of subsets of T. $\mathcal{G}$ $(\mathcal{H})$ is a space of functions defined on T with values in $Y\,(Z)$. Finally $Q=graph\mathcal{P}\subset\mathcal{G}\times\mathcal{H}$. Corollary 10 page 115 now gives a representation theorem for an independent relation on $\mathcal{G}\times\mathcal{H}$.

Theorem 66 *Let $\mathcal{G}\times\mathcal{H}$ be a mixture with respect to $\mathcal{A}$ where $\mathcal{A}$ has more than three disjoint non-null sets, Let $Q=graph\mathcal{P}$ be independent, open and essential for all partitions $(A_i)_{i\in I},(A_i\in\mathcal{A})$ of T and assume $(\mathcal{G}\times\mathcal{H})\,|A_i$ connected. Then there exists a function*

$$f:\mathcal{G}\times\mathcal{H}\times\mathcal{A}\rightarrow\mathbb{R}$$
such that
$$f(g,h,\cdot)\text{ is additive}$$
and
$$f(g,h,T)>0\Leftrightarrow(g,h)\in Q\Leftrightarrow g\in\mathcal{P}(h)$$

Corollary 21 *Let $\mathcal{G}$ be a mixture with respect to $\mathcal{A}$ where $\mathcal{A}$ has more than three disjoint non-null sets. Let $\mathcal{P}$ be a relation on $\mathcal{G}$ Let $graph\mathcal{P}=Q$ be independent, open and essential for all partitions $(A_i)_{i\in I},(A_i\in\mathcal{A})$ of T and assume $\mathcal{G}\,|A_i$ connected. Then there* *exists a function*

$$f:\mathcal{G}\times\mathcal{G}\times\mathcal{A}\to\mathbb{R}$$
such that
$$f(g,h,\cdot) \text{ is additive}$$
and
$$f(g,h,T)>0\Leftrightarrow(g,h)\in Q\Leftrightarrow g\in\mathcal{P}(h)$$

For $T=\{1,2,\cdots,n\}$ the notation $(y_1,y_2,\cdots,y_n),(z_1,z_2,\cdots,z_n)$ is used for the values of the functions g and h. Corollaries 10 and 11 reduce to

Corollary 22 *Let $\mathcal{P}$ be a relation on*

$$(Y_1\times Y_2\times\cdots\times Y_n)\times(Z_1\times Z_2\times\cdots\times Z_n)$$

Assume that $n>3$, that $\#(Y_i\times Z_i)>1$, that $\mathcal{P}$ is essential, independent and open, and that $(Y_i\times Z_i)$ is connected, then there exists a representation

$$f:(Y_1\times Y_2\times\cdots\times Y_n)\times(Z_1\times Z_2\times\cdots\times Z_n)\to\mathbb{R}$$
such that
$$f(y_1,y_2,\cdots,y_n,z_1,z_2,\cdots,z_n)=\sum w_t(y_t,z_t)$$
and
$$f(y,z)>0\Leftrightarrow y\in\mathcal{P}(z)$$

Theorem 67 *Let $\mathcal{P}$ be a relation on $(Y_1\times Y_2\times\cdots\times Y_n)$. Assume that $n>3$, that $\#Y_i>1$, that $\mathcal{P}$ is essential, independent and open, and that Y_i is connected, then there exists a representation*

$$f:(Y_1\times Y_2\times\cdots\times Y_n)\times(Y_1\times Y_2\times\cdots\times Y_n)\to\mathbb{R}$$
such that
$$f(y_1,y_2,\cdots,y_n,y_1',y_2',\cdots,y_n')=\sum w_t(y_t,y_t')$$
and
$$f(y,y')>0\Leftrightarrow y\in\mathcal{P}(y')$$

Remark 61 *If assumption A (local non-satiation) (page 43) holds, $w_t(y_t,y_t)=0$.*

The following special case for $\mathcal{G}$ - called the unit consumption space - is useful in the discussion of the acceptability of the independence assumption for preferences over time. It is from the mathematical point of view just another interpretation of subjective probability/uncertainty. (See section 8.8)

Example 14 *The special case* $T = \mathbb{R}_+, Y = Z = \{0, 1\}, (T, \mathcal{A}, \mathcal{P})$. *This is the special case where only consumption of* 0 *or* 1 *are possible, two alternative consumptions over time can therefore be represented by the subsets of* T, *where the consumption is* 1. *The relation* $\mathcal{P}$ *is defined on the class of indicator functions on* T, *or equivalently on* $\mathcal{A}$. *The assumption for a representation theorem are of course unchanged.*

Theorem 68 *Let* $(T, \mathcal{A}, \mathcal{P})$ *be a space, where* $(T, \mathcal{A})$ *has more than three disjoint subsets* A_i *for which* $A_i \in \mathcal{P}(\emptyset)$. *Let* $\mathcal{P}$ *be independent and open, let* $\mathcal{A}\,|A_i$ *be connected. Then there exist functions*

$$\alpha : \mathcal{A} \to \mathbb{R}_+ \text{ and } \beta : \mathcal{A} \to \mathbb{R}_+ \text{ additive}$$
$$\text{such that}$$
$$f(0, 0, A) = f(1, 1, A) = 0,$$
$$\alpha(A) = f(1, 0, A),\ \beta(A) = -f(0, 1, A)$$
$$\text{and}$$
$$\alpha(A \backslash B) - \beta(B \backslash A) > 0 \Leftrightarrow A \in \mathcal{P}(B)$$

Proof. See section 8.8 page 98. ■
Rewriting we get

$$\lambda(A) - \lambda(B) > \mu(A \triangle B) \Leftrightarrow A \in \mathcal{P}(B)$$
$$\text{where}$$
$$\alpha = \lambda - \mu,\ \beta = \lambda + \mu$$

where λ can be regarded as a discounting measure and μ as an uncertainty measure. Getting one unit in period A is then preferred to getting one unit in period B if the difference in the discounting measure is larger than the uncertainty. The uncertainty is measured as the value of the uncertainty measure on the symmetric difference of the two sets.

Two other ways of rewriting the representation theorem will be useful

$$A \in \mathcal{P}(B)$$
$$\Leftrightarrow$$
$$\lambda(A) - \lambda(B) > 0, \text{ for } \forall\lambda \text{ with } \alpha \leq \lambda \leq \beta$$
$$\Leftrightarrow$$
$$\lambda(A) - \lambda(B) > 0, \text{ for } \forall\lambda \text{ with } \lambda(T) = 1$$
$$\text{and } p(A) \leq \lambda(A) \leq q(A)$$
$$\text{where}$$
$$p(A) = \frac{\alpha(A)}{\alpha(A) + \beta(A^c)} \text{ and } q(A) = \frac{\beta(A)}{\beta(A) + \alpha(B^c)}$$

$p,q:\mathcal{A}\to[0,1]$ are non-additive set functions, but has of course very special properties[1]. A is then preferred to B if the discount measure of A is larger than the discount measure of B for all normalized measures between the lower and upper discount measures. All the different reformulations of the basic theorem will be useful.

Remark 62 *If $\mathcal{P}$ is continuous with respect to Lebesgue measure on T all the measures defined will also be continuous with respect to Lebesgue measure on T.*

17.2.1 Y general

The starting point is the integral representation in section 11.2

$$f(g,h,A)=\int_A W(g(t),h(t),t)\,dt$$

This integral representation requires special assumptions (see section 11.2 page 152), The assumptions will hold for all or most applications.

Chose functions g_0 and g_1 such that $f(g_1,g_0,A)>0$ for all A for which such functions exist[2]. Define

$$\alpha:\mathcal{A}\to\mathbb{R}\text{ by }\alpha(\cdot)=f(g_1,g_0,\cdot)$$
and
$$\beta:\mathcal{A}\to\mathbb{R}\text{ by }\beta(\cdot)=-f(g_0,g_1,\cdot)$$

Then some trivial manipulation gives first

$$f(g,h,A)>0\Leftrightarrow\int_A w(g(t),h(t),t)\,d\lambda>0\text{ for }\alpha\leq\lambda\leq\beta$$

and then

$$\begin{array}{ll} f(g,h,A)= & \\ \int_A u(g(t),h(t),t)\,d\lambda & \text{(Expected pure preference} \\ - & - \\ \int_A v(g(t),h(t),t)\,d\lambda & \text{expected } Y\text{-uncertainty} \\ - & - \\ \int_A |w(g(t),h(t),t)|\,d\mu & \text{uncertainty on }\mathcal{A}) \end{array}$$

[1] $p(A)+q(A^c)=1$. In the terminology of mathematical programming $-p$ and q are fully submodular and compliant.

[2] A natural choice will be to require the v defined below to be as small as possible but positive.

where

$$w, u, v : Y^2 \times T \to \mathbb{R},$$
$$w = u - v = w^+ + w^-; w^+, -w^- \geqq 0$$
$$u(\cdot, \cdot, t) \text{ is skew symmetric.} \qquad \text{(Pure preference)}$$
$$v(\cdot, \cdot, t) \text{ is symmetric.} \qquad \text{(Uncertainty on } Y\text{)}$$
$$w^+ = \tfrac{dt}{d\alpha} W^+, w^- = \tfrac{dt}{d\beta} W^-$$
$$\text{and}$$
$$\alpha = \lambda - \mu, \beta = \lambda + \mu$$

A function g is now preferred to another function h if the expected pure preference - $\int_T u d\lambda$ - is larger than the sum of the two types of uncertainty. First the expected value - $\int_T v d\lambda$ - of the uncertainty on Y with respect to the same discounting measure, and secondly the uncertainty on $\mathcal{A}$ weighted by the absolute value of the preference - $\int_T |w| \, d\mu$.

This is what a direct application of the uncertainty representation theorems gives. At each point in time there is a preference function, when this preference function is split into pure preference and uncertainty, the decomposition also gives uncertainty about the time preference - i.e. the weight of the preference at time t in the overall preference - there is a lower weight used when the preference is positive and a larger weight used when the preference is negative. One way of expressing this to say that the time preference expresses how fast subjective time goes. If the preference is negative subjective time moves slowly and the negative preference gets a large weight. If the preference is positive time moves faster, and the positive preference gets a smaller weight.

With this way of expressing the consequences of the independence assumption on the set of all functions $\mathcal{G}$, it becomes doubtful if it should be accepted. It can be argued that the uncertainty at time t which in this formulation only give rise to an interval of weights for preferences at time t also should be uncertainty about the weight on the preferences from time t and the rest of time. In order to allow for this possibility, the independence assumption may be changed. This idea will be sketched in the next section.

17.3 Existence and decomposition of $f : \mathcal{G} \times \mathcal{H}\times \dot{\mathcal{G}} \times \dot{\mathcal{H}} \times\mathcal{A} \to \mathbb{R}$

Let again $T = \mathbb{R}_+$ (in the interpretation time), Y and Z are arbitrary sets such that differentiation can be defined and $\mathcal{A}$ a system of subsets of T. $\mathcal{G}$ $(\mathcal{H})$ is a space of functions defined on T with values in Y (Z)

Another way of relaxing the independence assumption may be of interest. For $T = \mathbb{R}_+$ the functions $\mathcal{G} \times \mathcal{H}$ may be differentiable with derivatives

$$\dot{\mathcal{G}} \times \dot{\mathcal{H}} = \frac{d}{dt} (\mathcal{G} \times \mathcal{H})$$

Finally

$$Q = graph\mathcal{P} \subset \mathcal{M}\left(\mathcal{G} \times \mathcal{H}\times \dot{\mathcal{G}} \times\dot{\mathcal{H}}\right) \times \mathcal{M}\left(\mathcal{G} \times \mathcal{H}\times \dot{\mathcal{G}} \times\dot{\mathcal{H}}\right)$$

Where

$$\mathcal{M}\left(\mathcal{G} \times \mathcal{H}\times \dot{\mathcal{G}} \times \dot{\mathcal{H}} \right)$$

is the smallest $\sigma-$finite mixture of $\mathcal{G} \times \mathcal{H}\times \dot{\mathcal{G}} \times \dot{\mathcal{H}}$. (See definition 99 and remark 28 page 86). Corollary 10 page 115 now gives a representation theorem for the restriction of the relation $\mathcal{P}$ to $\mathcal{G}\times\mathcal{H}$ (and for the independent relation on $\mathcal{G}\times\mathcal{H}\times \dot{\mathcal{G}} \times \dot{\mathcal{H}}$). The notation $g \in \mathcal{P}(h)$ is used for $\left(g, \dot{g}\right) \in \mathcal{P}\left(h, \dot{h}\right)$ for $\left(\dot{g},\dot{h}\right) = \frac{d}{dt}(g, h)$

Theorem 69 *Let $\mathcal{A}$ have more than three disjoint non-null sets. Let $Q = graph\mathcal{P}$ be independent, open and essential for all partitions $(A_i)_{i\in I}, (A_i \in \mathcal{A})$ of T and assume $\mathcal{M}\left(\mathcal{G} \times \mathcal{H}\times \dot{\mathcal{G}} \times \dot{\mathcal{H}}\right)|A_i$ connected. Then there exists a function*

$$f : \mathcal{G} \times \mathcal{H}\times \dot{\mathcal{G}} \times \dot{\mathcal{H}} \times\mathcal{A} \to \mathbb{R}$$

such that

$$f\left(g, h, \dot{g}, \dot{h}, \cdot\right) \text{ is additive}$$

and

$$f\left(g, h, \dot{g}, \dot{h}, T\right) > 0 \Leftrightarrow \left(g, h, \dot{g}, \dot{h}\right) \in Q \Leftrightarrow g \in \mathcal{P}(h)$$

Proof. A reinterpretation of theorem 10. ■

Corollary 23 *Let $\mathcal{A}$ has more than three disjoint non-null sets. Let $\mathcal{P}$ be a relation on $\mathcal{M}\left(\mathcal{G}\times\dot{\mathcal{G}}\right)$. Let graph$\mathcal{P} = Q$ be independent, open and essential for all partitions $(A_i)_{i\in I}, (A_i \in \mathcal{A})$ of T and assume $\mathcal{M}\left(\mathcal{G}\times\dot{\mathcal{G}}\right)|A_i$ connected. Then there exists a function*

$$f : \mathcal{G}\times\mathcal{G}\times\dot{\mathcal{G}}\times\dot{\mathcal{G}}\times\mathcal{A}\rightarrow\mathbb{R}$$

such that

$$f\left(g, h, \dot{g}, \dot{h}, \cdot\right) \text{ is additive}$$

and

$$f\left(g, h, \dot{g}, \dot{h}, T\right) > 0 \Leftrightarrow \left(g, h, \dot{g}, \dot{h}\right) \in Q \Leftrightarrow g \in \mathcal{P}(h)$$

Proof. Put $\mathcal{H} = \mathcal{G}$ in theorem 69. ■

The following decomposition may be important for formulations of uncertainty in preferences over time

Definition 163 (Decomposition) *Define f_0 and f_1 by*

$$f_0(g, h, \cdot) = f(g, h, 0, 0, \cdot)$$
$$f_1\left(g, h, \dot{g}, \dot{h}, \cdot\right) = f\left(g, h, \dot{g}, \dot{h}, \cdot\right) - f(g, h, 0, 0, \cdot)$$

so

$$f\left(g, h, \dot{g}, \dot{h}, \cdot\right) = f_0(g, h, \cdot) + f_1\left(g, h, \dot{g}, \dot{h}, \cdot\right)$$

or

$$f = f_0 + f_1$$

The extreme cases $f = f_0$ or $f = f_1$ are both possible.

If the conditions for integral representation (see page 194) hold we get

$$f_0(g, h, T) =$$
$$\int_T u_0\left(g(t), h(t), \dot{g}(t), \dot{h}(t), t\right) d\lambda_0 \qquad \text{(Exp. pure preference}$$
$$-\int_T v_0\left(g(t), h(t), \dot{g}(t), \dot{h}(t), t\right) d\lambda_0 \qquad \text{- exp. } Y\text{-uncertainty}$$
$$-\int_T \left|w_0\left(g(t), h(t), \dot{g}(t), \dot{h}(t), t\right)\right| d\mu_0 \qquad \text{- uncertainty on } \mathcal{A})$$

and

$$
\begin{array}{ll}
f_1(g,h,X) = & \\
\left.\begin{array}{l} U_1(g(0),h(0),0)+ \\ \int_T u_1\left(g(t),h(t),\dot{g}(t),\dot{h}(t),t\right) d\lambda_1 \end{array}\right\} & \text{(Exp. pure preference} \\
\left.\begin{array}{l} -V_1(g(0),h(0),0) \\ -\int_T v_1\left(g(t),h(t),\dot{g}(t),\dot{h}(t),t\right) d\lambda_1 \end{array}\right\} & \text{- exp. } Y\text{-uncertainty} \\
-\int_T \left| w_1\left(g(t),h(t),\dot{g}(t),\dot{h}(t),t\right)\right| d\mu_1 & \text{- uncertainty on } \mathcal{A})
\end{array}
$$

The interpretation of the preferences and the uncertainties are as follows.

The extreme case $f = f_0$ is just the special case covered by the last section. The pure preference u_0 is the preference for g over h at point t in time. The $Y-$uncertainty v_0 is the corresponding uncertainty in the preferences for consumption at time t. The time preference λ_0 is the weight of preferences at time t. The uncertainty μ_0 is the uncertainty about the weight of the preferences at time t. All of these functions and measures are defined for a $t \in T$. (The measures can be defined by their derivatives with respect to Lebesgue measure on T).

The other extreme case $f = f_1$ is the case where the uncertainty at a time t is uncertainty also about all the future. $U_1(g(0),h(0),0)$ is the pure preference in the choice between $g(0)$ and $h(0)$ as constant consumptions starting from time 0 $V_1(g(0),h(0),0)$ is the uncertainty between $g(0)$ and $h(0)$ as constant consumptions starting from time 0. $U_1(g(0),h(0),0)$ and $V_1(g(0),h(0),0)$ will be 0. for $g(0) = h(0)$. u_1 is the marginal pure preference for g over h assuming that $(g(t),h(t))$ is the constant consumption from time t to the end of time. The measure λ_1 is the weight on this preference. v_1 is the corresponding uncertainty. μ_1 is the uncertainty weight on the preferences at time t for consumption from time t to the end of time. A special case would be that

$$u_1\left(g(t),h(t),\dot{g}(t),\dot{h}(t),t\right) = \frac{dU_1}{dy}\dot{g}(t) + \frac{dU_1}{dz}\dot{h}(t)$$

and

$$v_1\left(g(t),h(t),\dot{g}(t),\dot{h}(t),t\right) = \frac{dV_1}{dy}\dot{g}(t) + \frac{dV_1}{dz}\dot{h}(t)$$

where U_1 and V_1 are the pure preference and the uncertainty in choice at time t between the two constant consumptions $g(t), h(t)$

In general both f_0 and f_1 will be different from 0 and all four types of uncertainty can appear.

17.4 Notes

1. The most important problem not covered by any of the results is how the preferences at time t for consumption at time t and in the future is related to the preferences at time 0. It is an obviously unacceptable assumption that the preferences are just the conditional preferences from time 0. Both (or all four) types of uncertainty will have changed or may even have disappeared. Elegant formulations corresponding to stationarity would be nice to have.

2. Equally important will be to use the different representations of preferences over time to obtain characterizations of optimal controls and optimal trajectories (See Pontryagin et al. (1962) [149]). The result may be that there are many optimal choices and that there among the many optimal controls will be piecewise constant solutions, thus confirming Bewley's ideas. (See [21] and [24] and Rigotti Shannon (2001) [156]).

3. There is a large literature on preferences over time and preferences over time and uncertainty. Both theoretical and empirical studies suggest that the independence condition giving additivity over time and expected utilities is too strong. There may be habit formation - where the preferences now are not independent of the past - or recursive preferences - where the preferences for the future is independent of the past, but not the full independence condition. On the other hand there are results assuming that the preferences are translation invariant (see page 178). There has been papers including "Knightian uncertainty" or "ambiguity". The results in this chapter is only an indication of the type of results which can be obtained with uncertainty as formalized in this book. It will be very important for many parts of economic theory to include uncertainty about the preferences and the knowledge about

the future in a tractable way. The results and techniques from this book should be useful also for the more realistic representations.

18

A foundation for statistics

18.1 Introduction and historical background

Probability theory has had an almost unquestioned foundation since Kolmogorov (1933) [111]. Probability is a (normalized) measure on an algebra of events i.e. subsets of an arbitrary set. The probability of an event may be a result in a theory about any part of the real world. It may be an assumption that the beliefs or knowledge of agents can be expressed as total preorders on a system of subsets of events. Theorem 9 page 98 then gives probability as a representation of this relation. Under the assumptions of this representation theorem the two assumptions - a total preorder on an algebra or a probability measure on this algebra - are therefore equivalent.

The foundation of statistics was discussed by among others Fisher, Keynes and Jeffreys around 1920. Fisher (1922) [77] described the role of statistics as the "reduction of data" (page 311 [77]). This reduction of data "is accomplished by constructing a hypothetical infinite population, $\cdots$. The laws of distributions of this hypothetical population is specified by relatively few parameters, $\cdots$." (page 311 [77]). (See also Jeffreys (1961) [100]). Keynes (1921) [106] stressed the importance of a concept of uncertainty, which could not be expressed by probabilities. His views did not have any influence on the development of statistics. Neymann Pearson (1936,1938) [142] and many others developed methods of estimation and testing of hypothesis in this framework. Breiman (2001) [35] has recommended a much broader view on the role of statistics.

Savage (1954) [160] elaborated the Bayesian approach that the representation theorem should be used to give a probability measure on the space of parameters. The role of observations is then to give conditional measures on for example the parameter space. Lindley (1990) [125] and many others believe, that this approach is the only "coherent" approach to a foundation of statistics. Lehmann (1990) [121] and many others criticized the assumption that a total preorder on a system of subsets of the parameter space is given and known. To quote Lehmann (1959) [120] page 13: "This assumption is usually not warranted in applications"

Good (1950) [86], Dempster (1967 and 1968) [57, 58] and Berger (1984) [18] started the development of "robust" statistics, where the assumption that there is one probability measure on the parameter space, is replaced by the assumption that there is a class of probability measures on the parameter space. Berger Berlinger (1986) [19], de Robertis Hartigan (1981) [48], Wasserman (1990a,b) [185, 186], Wasserman Kadane (1992) [187] and Bose (1994) [32] are among many later papers developing this idea.

The purpose of this chapter is to sketch a foundation for statistics in terms of uncertainty as developed in this book. The idea in this foundation for statistics can be expressed simply by saying that the in both of the roles of probability measures in the Bayesian approach (A probability measure on the parameter space, and the parameter space is a space of probability measures) the probability measures are replaced by uncertainty in the form of convex sets of probability measures. Not all convex sets of probability measures on the parameter space will appear, and one parameter in the parameter space is no longer a probability measure but a set of probability measures on the sample space. Also in this case not all convex sets of probability measures will appear. Both convex sets are determined by intervals of measures on sub-$\sigma-$algebras. In that way also the uncertainty about whether the parameter space itself is the right space can be formalized.

18.2 Basic concepts

The basic concepts for statistics are a sample space $(\Omega, \mathcal{A})$, where Ω is an arbitrary set and $\mathcal{A}$ is a $\sigma-$algebra on Ω, and a parameter space Θ, where $\theta \in \Theta$ is a probability measure on $(\Omega, \mathcal{A})$. For Bayesian statisticians is furthermore a measure λ on $(\Theta, \mathcal{B})$ either given or obtained from a representation of a total preorder on the σ-algebra $\mathcal{B}$ of subsets of Θ.

To get started on Bayesian statistics some form of the following theorem is needed

Theorem 70 *Assume that all $\theta \in \Theta$ has a density $f : \Omega \times \Theta \rightarrow \mathbb{R}$ with respect to a $\sigma-$finite measure ν on $(\Omega, \mathcal{A})$, where f is measurable with respect to $\mathcal{A} \otimes \mathcal{B}$.*
Then there exists a unique measure γ on $(\Omega \times \Theta, \mathcal{A} \otimes \mathcal{B})$, such that the marginal measure on $(\Theta, \mathcal{B})$ is λ and conditional measure given $\theta \in \Theta$ is $\mathcal{B}-$a.e. θ. The mapping determined by this theorem $(\theta, \lambda) \longmapsto \gamma$ is thus a $\mathcal{B}-$a.e. bijection.

Proof. See for example de Robertis Hartigan (1981) page 238 [48] ∎

The relations between the different concepts are

$$\theta(A) = \int_A f(x, \theta) \, d\nu \in [0, 1] \qquad \text{(Probability on } \Omega \text{ given } \theta)$$

$$\lambda(B) = \gamma(B \times \Omega) \in [0, 1] \qquad \text{(Prior measure on } \Theta)$$

$$\theta = \frac{d\gamma}{d\lambda} : \mathcal{A} \rightarrow [0, 1] \qquad \text{(Conditional measure on } \Omega)$$

The prior probability of $A \in \mathcal{A}$ is

$$\int_{\theta \in \Theta} \theta(A) \, d\lambda$$

It is thus equivalent to assume that there is a total preorder on $\mathcal{B}$ (which can be represented by a measure λ) and to assume that there is a total preorder on $\mathcal{A} \otimes \mathcal{B}$ (which can be represented by a measure γ). The nicest abstract formulation of Bayesian statistics is therefore obtained by starting with a sample space and a parameter space and a σ-algebra on the product and then assume the existence of a total preorder on the σ-algebra. Then the representation theorem can be used to give a measure γ on the product. This measure gives as the marginal measure on $(\Theta, \mathcal{B})$ the prior measure λ, and the conditional measures on $(\Omega, \mathcal{A})$ given $\theta \in \Theta$ gives the parameter space as a set of probability measures on the sample space $(\Omega, \mathcal{A})$.

Definition 164 (Random variables, observations) *Random variables are now measurable functions ξ, η defined on Ω with values in measurable spaces X, Y. Values of a random variables $(x, y) = (\xi, \eta)(\omega)$ are called observations. So*

$$(x, y) \in X \times Y \text{ and } (\xi, \eta) \in (X \times Y)^{\Omega}$$

Remark 63 *A value of a parameter $\theta \in \Theta$ determines a probability distribution for any random variable and for any sub $\sigma-$algebra $\mathcal{A}_0$ of the given algebra $\mathcal{A}$. This also means that given an observation x the parameter and the observation determine conditional distributions on further observations y and determines conditional measures on $\mathcal{A}$ (and on $\mathcal{A}_1 \subset \mathcal{A}$) given $\mathcal{A}_0 \subset \mathcal{A}$. If the parameter has independence between ξ and η, the distribution of η is independent of x, but depends of course on the parameter.*

Given a probability measure λ on the parameter space and an observation x of a random variable ξ with the density $h(x, \theta)$ the posterior distribution on the parameter space has the density

$$l(x, \theta) = \frac{h(x, \theta)}{\int h(x, \theta)\, d\lambda} \in \mathbb{R}_+$$

and the conditional probability on $\mathcal{A}$ is given by

$$\theta(A, x) = \int_A l(y, \theta)\, d\nu \in \mathbb{R}_+$$

This density function on the parameter space and this probability on $\mathcal{A}$ is then for Bayesian statisticians what the observation x says about the parameter space and about probabilities on further observations.

For others $h(x, \cdot)$ is what the observation x says about the parameter space, and they may replace the parameter space with the $\arg\max_{\theta \in \Theta} l(x, \theta)$ the maximum likelihood estimate for the parameter and in that way be able to get one probability measure on further observations. Or they more generally define functions on the observations with values in the parameter space or in the set of for example intervals in the parameter space, and in that way use the observations to get probability measures on further observations.

Common for both classes of statisticians is, that they have to chose a parameter space[1]. The difference is that some in addition are willing to specify a total preorder on subsets of the parameter space and therefore a probability measure on the parameter space. Others maintain that this is impossible or arbitrary.

18.3 Uncertainty about the parameter space

There are two forms of uncertainty about the parameter space.

It may be uncertain where in the parameter space the right probability distribution is. Bayesians are willing to express this uncertainty by probabilities, others are not.

It may be uncertain if the right probability distribution is in the parameter space. This is sometimes called model risk, see for example Cairns (2000) [36]. Neither Bayesian statisticians or other statisticians have a nice abstract formulation of this uncertainty.

18.4 Robust Bayesian inference

Many Bayesians have accepted the critique of the assumption that there is a total preorder on the subsets of a parameter space, and tried to answer it by replacing one probability measure on the parameter space by a class of measures or by more general set functions than measures. See the references in section 18.1

The theory does not have a nice axiomatic foundation, and it can only be used to express the first kind of uncertainty.

18.5 Requirements for a foundation of statistics

1. Knowledge and uncertainty should enter in the same way as in other sciences.

2. Both forms of uncertainty should be expressed.

3. Observations should influence both forms of uncertainty, so observations should also increase or decrease the uncertainty

[1] See however [35]

about whether the parameter space contain the right probability measure.

4. The way uncertainty is expressed (a measure, a class of measures, a belief function etc.) should be the same before and after an observation.

5. The expressions for knowledge and uncertainty appearing in the theory should be computable.

18.6 A foundation of statistics

A foundation of statistics which fulfills the requirements is now obtained by regarding statistics as a two level uncertainty problem. The basic concepts are the same as above. A sample space $(\Omega, \mathcal{A})$, where Ω is an arbitrary set and $\mathcal{A}$ is a $\sigma-$algebra on Ω, and a parameter space Θ, where $\theta \in \Theta$ is a set of measures on $(\Omega, \mathcal{A})$. Given a value of the parameter there is uncertainty on functions on the sample space, and on the parameter space there is uncertainty again expressed as a set of measures. So remark 63 can be repeated:

Remark 64 *A value of a parameter $\theta \in \Theta$ determines uncertainty for any random variable and for any sub $\sigma-$algebra $\mathcal{A}_0$ of the given algebra $\mathcal{A}$. The parameter also determines conditional uncertainty on further observations y given an observation x of one random variable ξ and determines conditional uncertainty on $\mathcal{A}$ (and on $\mathcal{A}_1 \subset \mathcal{A}$) given $\mathcal{A}_0 \subset \mathcal{A}$*

More explicitly: Given $\theta \in \Theta$ there is a $\mathcal{P}_\theta$ on $\mathcal{A}$ (or on a set of measurable functions on $(\Omega, \mathcal{A}))$[2] represented by a set of measures $(\Sigma_{\theta i})_{i \in I}$ and a set of probability measures Λ_θ on $\mathcal{A}$ (and a preference function on $\Omega \times Y \times Y$) determined as in chapters 9 and 14 i.e.

$$\begin{aligned} \Sigma_{\theta i} &= \left\{\mu : \mathcal{A} \to \mathbb{R}_+ \left| \alpha_i^\theta \leqq \mu \right| A_i \leqq \beta_i^\theta \right\}, i \in I_\theta \\ \Lambda_\theta &= \{\mu : \mathcal{A} \to \mathbb{R}_+ \left| p_i \leqq \mu \right| A_i \leqq b_i \; i \in I_\theta\} \end{aligned}$$

[2] The generalization to this case follows as in the earlier chapters, measures have to be replaced by expected values of preference functions.

where

$$p_i(A) = \frac{\alpha_i^\theta(A)}{\alpha_i^\theta(A)+\beta_i^\theta(A^c)} \text{ and } b_i(A) = \frac{\beta_i^\theta(A)}{\beta_i^\theta(A)+\alpha_i^\theta(A^c)}$$
$$\text{for } A \in \mathcal{A}_i$$

and

$$\alpha_i^\theta, \beta_i^\theta : \mathcal{A}_i \to \mathbb{R}_+, \ \sigma\text{-additive, for } i \in I_\theta$$
$$\alpha_i^\theta(A \setminus A') > \beta_i^\theta(A' \setminus A) \Longleftrightarrow A \in \mathcal{P}_i(A'), A, A' \in \mathcal{A}_i$$

and finally the consistent extension $\mathcal{P}_\theta$ of $(\mathcal{P}_{\theta i})_{i\in I}$

$$A \in \mathcal{P}_\theta(A') \Longleftrightarrow \mu(A) > \mu(A') \ \text{ for all } \mu$$

On the parameter space there is also uncertainty $\mathcal{P}$ represented by

$$\Sigma_j = \left\{\lambda : \mathcal{B} \to \mathbb{R}_+ \left|\alpha_j^\theta \leqq \lambda\right|\mathcal{B}_j \leqq \beta_j^\theta\right\}, j \in J$$
$$\Lambda_\Theta = \left\{\lambda : \mathcal{B} \to [0,1] \left|p_j \leqq \lambda\right|\mathcal{B}_j \leqq b_j \ j \in J\right\}$$

where

$$p_j(B) = \frac{\alpha_i^\theta(B)}{\alpha_i^\theta(B)+\beta_i^\theta(B^c)} \text{ and } b_i(B) = \frac{\beta_i^\theta(B)}{\beta_i^\theta(B)+\alpha_i^\theta(B^c)}$$
$$\text{for } B \in \mathcal{B}_j$$

and

$$\alpha_j^\theta, \beta_j^\theta : \mathcal{B}_j \to \mathbb{R}_+, \ \sigma\text{-additive, for } j \in J$$
$$\alpha_j^\theta(B \setminus B') > \beta_j^\theta(B' \setminus B) \Longleftrightarrow B \in \mathcal{P}_j(B'), B, B' \in \mathcal{B}_i$$

and the consistent extension $\mathcal{P}_\Theta$ of $(\mathcal{P}_j)_{j\in J}$

$$B \in \mathcal{P}_\Theta(B') \Longleftrightarrow \lambda(B) > \lambda(B') \ \text{ for all } \lambda \in \Lambda_\Theta$$

Now the measures from $(\Sigma_{\theta i})_{\theta, i\in\Theta\times I}$ can for any $\lambda \in \Lambda_\Theta$ following theorem 70 and (assuming that all measures from $(\Sigma_{\theta i})_{\theta, i\in\Theta\times I}$ have densities with respect to a $\sigma-$finite measure ν on $(\Omega, \mathcal{A})$, where the densities are measurable with respect to $\mathcal{A} \otimes \mathcal{B}$) be combined to measures on $(\Omega \times \Theta, \mathcal{A} \otimes \mathcal{B})$. The conditional measures with respect to θ will be denoted $\lambda(\cdot;\theta)$

The set of normalized measures on $(\Omega \times \Theta, \mathcal{A} \otimes \mathcal{B})$ for which both sets of conditions hold will be denoted Λ so

Definition 165 (a priori knowledge)

$$\begin{array}{c}\Lambda=\{\lambda:\mathcal{A}\otimes\mathcal{B}\rightarrow\mathbb{R}_{+}\,|\lambda\left(\cdot;\theta\right)\in\Lambda_{\theta},\lambda\left(\Omega,\cdot\right)\in\Lambda_{\Theta}\}\\ C'\in\mathcal{P}\left(C\right)\Longleftrightarrow\lambda\left(C'\right)>\lambda\left(C\right)\ \textit{for all}\ \lambda\in\Lambda\end{array}\tag{18.1}$$

Λ and $\mathcal{P}$ are determined by $\left(\alpha_i^\theta,\beta_i^\theta\right)_{i\in I_\theta}$ and $\left(\alpha_j^\theta,\beta_j^\theta\right)_{j\in J}$ but not conversely.

An observation x then give conditional normalized measures $\lambda\left(\cdot;x\right)\in\Lambda_x$ on $(\Omega\times\Theta,\ \mathcal{A}\otimes\mathcal{B})$ where

$$\begin{array}{c}\Lambda_x=\left\{\lambda\left(\cdot;x\right):\mathcal{A}\otimes\mathcal{B}\rightarrow\mathbb{R}_{+}\left|\lambda\left(\cdot\right)=\int\lambda\left(\cdot;x\right)d\lambda,\lambda\in\Lambda\right.\right\}\\ C'\in\mathcal{P}_x\left(C\right)\Longleftrightarrow\lambda\left(C'\right)>\lambda\left(C\right)\ \text{for all}\ \lambda\in\Lambda_x\end{array}$$

Λ_x and $\mathcal{P}_x$ answer the questions statisticians can be expected to answer: What is the uncertainty on not already observed random variables (has the uncertainty been reduced to probability, so the parameter space has been accepted as the right parameter space, or has the uncertainty become so large that the chosen statistical model has been rejected?)

It will require special assumptions to be able to express Λ_x and $\mathcal{P}_x$ in the form 18.1. So in general the importance of observations can not be reduced to changes in the uncertainty on the parameter space.

The answers given by statisticians will be somewhat different in form given this foundation for statistics. Statisticians need no longer accept or reject hypothesis. They need no longer give point or interval estimates of parameters. They will be able to give precise statements about probabilities and uncertainties for future events or unobserved random variables. For some observations the uncertainty may be so large that no useful statements can be made; for other observations the uncertainty may have almost disappeared. All conclusions will obviously still depend on the original choice of parameter space, no conclusions can be reached based on observations alone, but the conclusions reached will be what Lindley (1990) [125] called coherent.

There are two important differences between this foundation for statistics and the foundation being accepted by many Bayesians (Wasserman, etc.)

Firstly Wasserman and others accept very large classes of measures as knowledge about the parameter space (belief functions etc.). It is difficult to justify these classes axiomatically. In our foundation all forms of knowledge are expressed by pairs of measures on sub $\sigma-$algebras and the convex sets of probability measures generated by such sets of pairs of measures, and they can viewed as representations of independent relations on function spaces.

Secondly the parameter space for robust Bayesians is a set of probability measures on the sample space. Bayesians and other statisticians have the common problem, that they lack a consistent way of expressing uncertainty about the choice of parameter space, and of expressing that the role of observations is not just to determine where in the parameter space the right parameter is, but also to determine if the parameter space is the right space. In our foundation the parameter space is a convex set of measures on $\mathcal{A}$. For some subset of Ω the representation may mean that there is a probability measure and no uncertainty, for other parts of Ω there may be only uncertainty. This opens for the possibility that observations influence the uncertainty on further observations. Given the prior knowledge the set of measures conditional on the observations will also answer this question.

None of the five requirements to a foundation give serious problems.

1. The idea of a parameter space is central for statistics, and it is only used to a limited extend in theories about behavior under uncertainty. The way the uncertainty is formalized on the outcome space and on the parameter space should however be the same as in other sciences using uncertainty. (If the ideas and results developed in this book are useful also in other applications).

2. and

3. This is accomplished by having the parameter space be a space of uncertainty measures on the outcome space and not as usual a space of probability measures.

4. The conditional measures obtained after an observation will of course be different from the original sets of measures, but the form of the representations will be exactly the same. It is worth noticing that due to the properties of conditional

measures the conditional measures given an observation x_1, will contain all the relevant knowledge in the sense that the conditional measures given (x_1, x_2), will be the same as the conditional measures based on first x_1 and then x_2.

The past observations will not have to remembered when the conditional measures given x_1 is known.

5. Of course depends on how complicated the uncertainties and the parameter space are formulated.

18.7 Notes

The idea to use sets of measures on a parameter space is of course known from Neo-Bayesian theory. The result that only the assumption that the relation on subsets of the parameter space is total has to be changed and that the resulting representation will give very special sets of measures as representations is new. The idea to replace the parameter space as a space of probability measures on the outcome space with a space of sets of uncertainty measures may also be new[3].

[3] Similar ideas can be found in de Cooman (2002) [47].

References

[1] J. Aczél. The notion of mean values. *Norske Vidcnskabernes Selskabs Forhandlinger, Trondheim*, 19:83–86, 1946.

[2] J. Aczél. On mean values and operations defined for two variables. *Norske Videnskabernes Selskabs Forhandlinger, Trondheim*, 20:37–40, 1947.

[3] J. Aczél. On mean values. *Bull. Math. Soc.*, 54:392–400, 1948.

[4] J. Aczél. Quasigroups - nets - nomograms. *Advan. Math.*, 1:383–450, 1965.

[5] J. Aczél. *Lectures on Functional Equations and their Applications*, volume 19 of *Mathematics in Science and Engeneering*. Academic Press, New York, 1966.

[6] J. Aczél, G. Pickert, and F. Radó. Nomogramme, Gewebe und Quasigruppen. *Mathématica (Cluj)*, 2 (25):5–24, 1960.

[7] E.M. Alfsen. Order theoretic foundation of integration. *Math. Annalen*, 149:419–61, 1963.

[8] M. Allais. Fondement d'une théorie positive des choix comportant un risque et critique des postulats et axiomes de l'école américaine. *Econometrie*, 15:257–332, 1952.

[9] M. Allais. Le comportement de l'homme rationnel devant le risque: Critique des postulats et axiomes de l'école américaine. *Econometrica*, 21(4):503–46, October 1953.

[10] K.J. Arrow. Alternative approaches to the theory of choice in risk-taking situations. *Econometrica*, 19(4):404–37, October 1951.

[11] R.J. Aumann. Utility theory without the completeness axiom. *Econometrica*, 30(3):445–62, July 1962.

[12] R.J. Aumann. Measurable utility and the measurable choice theorem. *La Décision, Colloque Internationaux du C.N.R.S., Paris.*, pages 15–26., 1969.

[13] R. Baer. Nets and groups. *Transactions of the American Mathematical Society*, 46(1):110–141, July 1939.

[14] R. Baer. Nets and groups. II. *Transactions of the American Mathematical Society*, 47(3):435–39, May 1940.

[15] R. Baer. The fundamental theorems of elementary geometry. An axiomatic analysis. *Transactions of the American Mathematical Society*, 56(1):94–129, July 1944.

[16] S. Barbera, P. Hammond, and C. Seidl, editors. *Handbook of Utility Theory, Principles*, volume 1. Kluwer, Dordrecht, 1998.

[17] A.F. Beardon, J.C. Candeal, G. Herden, E. Induráin, and G.B. Metha. The non-existence of a utility function and the structure of non-representable preference relations. *Journal of Mathematical Economics*, 37:17–38, 2002.

[18] J. Berger. *The Robust Bayesian Viewpoint (with Discussion), in Robustness in Bayesian Analysis, J. Kadane Ed.*, pages 63–144. North- Holland, 1984.

[19] J. Berger and L.M. Berliner. Robust Bayes and empirical Bayes analysis with ε-contaminated priors. *The Annals of Statistics*, 14(2):461–86, 1986.

[20] D. Bernoulli. Specimen theoriae novae de mensura sortis. *Commentarii Academiae Scientiarium Imperiales Petropolitanae*, V:175–92, 1738. Translation "Exposition of a new theory on the measurement of risk" in Econometrica, 22 (1), January 1954 :23-36.

[21] T.F. Bewley. Knightian decision theory: Part 1. Cowles Foundation Discussion Paper 807, Cowles Foundation, Yale, November 1986. pages 52.

[22] T.F. Bewley. Knightian decision theory, part 2: Intertemporal problems. Cowles Foundation Discussion Paper 835, Cowles Foundation, Yale, May 1987.

[23] T.F. Bewley. Knightian decision theory and econometric inference. Cowlcs Foundation Discussion Paper 868, Cowles Foundation, Yale, March 1988.

[24] T.F. Bewley. *Market Innovation and Entrepreneurship: A Knightian View*, pages 41–58. Springer, 2001. In Festschrift for Werner Hildenbrand, edited by G. Debreu, W. Neuefeind, and W. Trockel.

[25] P. Billingsley. *Convergence of Probability Measures.* New York, 1968.

[26] G. Birkhoff. *Lattice Theory.* Providence, 1967.

[27] D. Blackwell and M.A. Girshick. *Theory of Games and Statistical Decisions.* New York, 1954.

[28] W. Blaschke. *Lectures on Topological Questions of Differential Geometry.* University of Chicago, Chicago, 1932.

[29] W. Blaschke. *Selected Problems of Differential Geometry.* Calcutta, 1934.

[30] W. Blaschke and G. Bol. *Geometrie der Gewebe.* Berlin, 1938.

[31] G. Bol. Über Kurvenscharen in Raum. *Abhandlungen aus dem Math. Sem. der Hamburgischen Univ.*, 7:399–405, 1930.

[32] S. Bose. Bayesian robustness with mixture classes of priors. *Annals of Statistics*, 22(2):652–67, 1994.

[33] N. Bourbaki. *Eléments de Mathématique.* Paris, 1939 -. (English Translation 1966 -).

[34] G. Bowen. A new proof of a theorem in utility theory. *International Economic Review*, 9:374, 1968.

[35] L Breiman. Statistical modeling: The two cultures. *Statistical Science*, 16(3):199–231, August 2001. Comments by D.R. Cox, B. Efron, B. Hoadley, and E. Parzen, and a Rejoinder by L. Breiman.

[36] A.J.G. Cairns. A discussion of parameter and model uncertainty in insurance. *Insurance: Mathematics and Economics*, 27:313–30, 2000.

[37] S.S. Chern. Abzälungen für Gewebe. *Abh. Math. Sem. Hamburg*, pages 163–70, 1935.

[38] S.S. Chern. Eine Invariantentheorie der Dreigewebe aus r-dimensionalen Mannigfaltigkeiten in R^{2r}. *Abh. Math. Sem. Hamburg*, pages 333–58, 1936.

[39] S.S. Chern. The geometry of g-structures. *Bull. Amer. Math. Soc.*, 72:167–219, 1966.

[40] S.S. Chern and P.A. Griffiths. Abel's theorem and webs. *Jahresberichte der Deutchen Matematiker Vereinigung*, pages 13–110, 1978.

[41] S.S. Chern and P.A. Griffiths. An inequality for the rank of a web and webs of maximum rank. *Ann. Scualo Norm. Sup. Pisa Cl. Sci.*, pages 539–57, 1978.

[42] S.S. Chern and P.A. Griffiths. Linearizations of webs of codimension one and maximum rank. In *Proc. Of International Symposium on Algebraic Geometry, Kyoto 1977*, 1978.

[43] J.L. Chipman, L. Hurwicz, M. Richter, and H. Sonnenschein, editors. *Preferences, Utility, and Demand.* Harcourt Brace Jovanovicch, New York, 1971.

[44] J.S. Chipman. The foundation of utility. *Econometrica*, 28(2):193–224, April 1960.

[45] H. Cramér. A theorem on ordered sets of probability distributions. *Theory of Probability and its Applications*, 1:16–25, 1956.

[46] D.B. Damiano. Webs and characteristic forms of Grassmann manifolds. *American Journal of Mathematics*, 105(6):1325–45, Dcember 1983.

[47] G. de Cooman. Precision-imprecision equivalence in a broad class of imprecise hierarchical uncertainty models. *Journal of Statistical Planning and Inference*, 105:175–98, 2002.

[48] L. de Robertis and J.A. Hartigan. Bayesian inference using intervals of measures. *Annals of Statistics*, 9(2):235–44, March 1981.

[49] G. Debreu. Une economie de l'incertain. (Chapter 8 in Mathematical Economics, Twenty Papers of Gérard Debreu (Debreu (1983)) contains a translation based on the paper), 1953.

[50] G. Debreu. *Representation of a Preference Ordering by a Numerical Function*, pages 159–65. in Decision Processes edited by Thrall et al., Wiley, New York, 1954. Chapter 6 in Mathematical Economics, Twenty papers by Gérard Debreu, Cambridge University Press 1983.

[51] G. Debreu. *Theory of Value, an axiomatic analysis of economic equilibrium.* Wiley, 1959.

[52] G. Debreu. *Topological Methods in Cardinal Utility Theory*, pages 16–26. Stanford University Press, 1960. In Mathematical Methods in the Social Sciences, 1959 edited by Arrow, Karlin and Suppes.

[53] G. Debreu. Continuity properties of Paretian utility. *International Economic Review*, 5:285–93, 1964.

[54] G. Debreu. Neighboring economic agents. *La Décision*, pages 85–90, 1969.

[55] G. Debreu. Smooth preferences. *Econometrica*, 40:603–14, 1972.

[56] G. Debreu. *Mathematical Economics, Twenty Papers of Gérard Debreu.* Cambridge, 1983.

[57] A. Dempster. Upper and lower probabilities induced from a multivalued mapping. *Annals of Mathematical Statistics*, 38:325–39, 1967.

[58] A. Dempster. A generalization of Bayesian inference. *Journal of the Royal Statistical Society, Series B*, 30:205–47, 1968.

[59] D. Denneberg. *Non-Additive Measure and Integral.* Kluwer, Dordrecht, 1994.

[60] D. Denneberg. Totally monotone core and products of monotone measures. *International Journal of Approximate Reasoning*, 24(2-3):273–81, 2000.

[61] P.A. Diamond. The evaluation of infinite utility streams. *Econometrica*, 33(1):170–77, January 1965.

[62] P.A. Diamond, T.C. Koopmans, and R.E. Williamson. Stationary utility and time perspective. *Econometrica, 32*, pages 82–100, 1964.

[63] D. Duffie and L.G. Epstein. Stochastic differential utility. *Econometrica*, 60(2):353–94, March 1992.

[64] N. Dunford and J. Schwartz. *Linear Operators Part I.* Interscience, 1958.

[65] B.E. Easton. Functional forms in economic theory. MRC Tecnical Report 1253, Mathematics Research Center, University of Wisconsin-Madison, September 1972.

[66] F.Y. Edgeworth. *Mathematical Psychics.* Paul Kegan, 1881.

[67] S. Eilenberg. Ordered topoplogical spaces. *American Journal of Mathematics*, 63(1):39–45, January 1941.

[68] D. Ellsberg. Risk, ambiguity, and the Savage axioms. *Quarterly Journal of Economics*, 75(4):643–669, November 1961.

[69] L.G. Epstein. The unimportance of intransitivity of separable preferences. *International Economic Review*, 28(2):315–22, June 1987.

[70] L.G. Epstein and J. Zhang. Subjective probabilities on subjectively unambiguous events. *Econometrica*, 69(2):265–306, March 2001.

[71] L.G. Epstein and S.E. Zin. Substitution, risk aversion, and the temporal behavior of consumption and asset returns: A theoretical framework. *Econometrica*, 57(4):937–69, July 1989.

[72] P. Fishburn and P. Wakker. The invention of the independence condition for preferences. *Management Science*, 41(7):1130–44, July 1995.

[73] P.C. Fishburn. Bounded expected utility. *Annals of Mathematical Statistics*, 38(4):1054–60, August 1967.

[74] P.C. Fishburn. Preference-based definitions of subjective probability. *Annals of Mathematical Statistics*, 38(6):1605–17, December 1967.

[75] P.C. Fishburn. Separation theorems and expected utilities. *Journal of Economic Theory*, 11(1):16–34, August 1975.

[76] P.C. Fishburn. Nontransitive measurable utility. *Journal of Mathematical Psychology*, 26:31–67, 1982.

[77] R.A. Fisher. On the mathematical foundation of theoretical statistics. *Philosofical Transactions of the Royal Society of London, Series A*, 222:309–68, 1922.

[78] A.A. Fraenkel. *Abstract Set Theory*. Amsterdam, 1961.

[79] A. Frank and E. Tardos. Generalized polymatroids and submodular flows. *Mathematical Programming*, 42:489–563, 1988.

[80] M. Friedman. *Price Theory: A Provisional Text*. Aldine, Chicago, 1976.

[81] L. Fuchs. On mean systems. *Acta Math.-Acad. Sci. Hun.*, 1:303–20, 1950.

[82] L. Fuchs. *Partially Ordered Algebraic Systems.* London, 1963.

[83] G. Fuhrken and M.K. Richter. Additive utility. *Economic Theory*, 1(1):83–105, November 1991.

[84] P. Ghirardato and M. Marinacci. Ambiguity made precise: A comparative foundation. *Journal of Economic Theory*, 102(2):252–89, February 2002.

[85] S.M. Goldman and H. Uzawa. A note on separability in demand analysis. *Econometrica*, 32(3):387–98, July 1964.

[86] I.J. Good. *Probability and the Weighing of Evidence.* Charles Griffin and Co., London, 1950.

[87] W.M. Gorman. The structure of utility functions. *Review of Economic Studies*, XXXV(4):367–90, October 1968.

[88] S. Grant, A. Kajii, and B. Polak. Decomposable choice under uncertainty. *Journal of Economic Theory*, 92(2):169–97, June 2000.

[89] P.A. Griffiths. Variations on a theorem of Abel. *Invetiones Mathematicae*, pages 321–90, 1976.

[90] P.A. Griffiths. On Abel's differential equation. *Algebraic Geometry*, pages 26–51, 1977.

[91] B. Grodal and J.-F. Mertens. Integral representation of utility functions. Technical Report CORE 6823, CORE, 1968.

[92] H. Hahn. Über die nichtarchimedischen Grössensysteme. *S.-B. Akad. Wiss. Wien*, 116:601–05, 1907.

[93] P.R. Halmos. *Measure Theory.* Princeton, 7 edition, 1961.

[94] F. Hausdorff. *Set Theory.* New York, 1967. First edition: Grundzüge der Mengenlehre, Leipzig 1914.

[95] I.N. Herstein and J. Milnor. An axiomatic approach to measurable utility. *Econometrica*, 21(2):291–97, April 1953.

[96] W. Hildenbrand. On economies with many agents. *Journal of Economic Theory*, 2:161–88, 1970.

[97] W. Hildenbrand. *Core and Equilibria of a Large Economy.* Princeton Studies in Mathematical Economics. Princeton University Press, Princeton, 1974.

[98] O. Hölder. Die Axiome der Quantität und die Lehre vom Mass. *Berichte über die Verhandlungen der Königlich Sächsischen Gesellschaft der Wisseschaften zu Leipzig, Math.-Phys. Classe*, 53:1–64, 1901.

[99] H.S. Houthakker. Additive preferences. *Econometrica*, 28:244–57, 1960.

[100] H. Jeffreys. *Theory of Probability.* Oxford University Press, Oxford, 3rd edition, 1961.

[101] D. Kahneman and A. Tversky. Prospect theory: An analysis of decisions under risk. *Econometrica*, 47(2):263–292, March 1979.

[102] E. Kamke. *Theory of Sets.* New York, 1950. First edition Mengenlehre Berlin 1928.

[103] E. Karni, D. Schmeidler, and K. Vind. On state dependent preferences and subjective probalities. *Econometrica*, 51(4):1021–31, July 1983.

[104] J.L. Kelley. *General Topology.* Princeton, 1955.

[105] J.L. Kelley, I. Namioka, and Co-Authors. *Linear Topological Spaces.* Princeton, 1963.

[106] J.M. Keynes. *A Treatise of Probability.* Mcmillan, 1921.

[107] J.M. Keynes. *The General Theory of Employment Interest and Money.* Macmillan, London, 1936.

[108] R. Kihlstrom, A. Mas-Colell, and H. Sonnenschein. The demand theory of the weak axiom of revealed preference. *Econometrica*, 44(5):971–78, 1976.

[109] T. Kim and M.K. Richter. Nontransitive-nontotal consumer theory. *Journal of Economic Theory*, 38(2):324–63, 1986.

[110] F.H. Knight. *Risk, Uncertainty, and Profit.* Houghton Mifflin, New York, 1921.

[111] A.N. Kolmogorov. *Grundbegriffe der Wahrscheinlichkeitsrechnung.* Springer, Berlin, 1933.

[112] A.N. Kolmogorov. On the representation of continuous functions of many variables by superposition of continuous functions of one varible and addition. *(Russian) Dokl. Akad. Nauk. SSSR*, 114:953–56, 1957. Translated (1963) Am. Math. Soc. Transl. 28, 55-59.

[113] B. Koopman. The axioms and algebra of intuitive probability. *Annals of Mathematics*, 41:269–92, 1940.

[114] T.C. Koopmans. Stationary ordinal utility and impatience. *Econometrica*, 28:287–309, 1960.

[115] T.C. Koopmans. Structure of preference over time. Cowles Foundation Discussuion Paper 206, Cowles Foundation, April 1966.

[116] T.C. Koopmans. Preference orderings over time. *Decision and Organization*, 1972. ed. by C.B. McGuire and R. Radner, Amsterdam.

[117] H.J. Kowalsky. *Topological Spaces.* New York, 1965.

[118] D.H. Krantz. Conjoint measurement: The Luce-Tukey axiomatization and some extensions. *Journal of Mathematical Psychology*, 1:248–77, 1964.

[119] D.M. Kreps and E.L. Porteus. Temporal resolution of uncertainty and dynamic choice theory. *Econometrica*, 46:185–200, 1978.

[120] E.L. Lehmann. *Testing Statistical Hypotheses.* Wiley, New York, 1959.

[121] E.L. Lehmann. The 1988 Wald memorial lectures: The present position in Bayesian statistics: Comment. *Statistical Science*, 5(1):82–83, February 1990.

[122] W.W. Leontief. Introduction to a theory of internal structure of functional relationships. *Econometrica*, 15:361–73, 1947.

[123] W.W. Leontief. A note on the interrelation of subsets of independent variables of a continuous with continuous first derivatives. *Bull. Amer. Math. Soc.*, 53:343–50, 1947.

[124] S.F. LeRoy and L.D. Singel. Knight on risk and uncertainty. *The Journal of Political Economy*, 95(2):394–406, April 1987.

[125] D.V. Lindley. The 1988 Wald memorial lectures: The present position in Bayesian statistics. *Statistical Science*, 5(1):44–65, February 1990.

[126] M. Loève. *Probability Theory.* van Norstrand, second edition, 1960.

[127] G. Loewenstein and D. Prelec. Anomalies in intertemporal choice: Evidence and an interpretation. *The Quarterly Journal of Economics*, 107(2):573–97, May 1992.

[128] G.C. Loomes and R. Sugden. Regret theory: An alternative theory of rationel choice under uncertainty. *Economic Journal*, 92:805–24, 1982.

[129] P. Lorenzen. Über halbgeordnete Gruppen. *Math. Zeits.*, 52:483–526, 1950.

[130] R.D. Luce and J.W. Tuckey. Simultaneous conjoint measurement: A new type of fundamental measurement. *Journal of Mathematical Psychology*, 1:1–27, 1964.

[131] M.J. Machina. Dynamic consistency and non-expected utility models of choice under uncertainty. *Journal of Economic Literature*, 27:1622–88, 1989.

[132] M.J. Machina and D. Schmeidler. A more robust definition of subjective probability. *Econometrica*, 60(4):745–80, July 1992.

[133] S. MacLane and G. Birkhoff. *Algebra*. Macmillan, New York, 1967.

[134] J. Marschak. Rational behaviour, uncertain prospects and measurable utility. *Econometrica*, 18:111–41, 1950.

[135] A. Mas-Colell. An equilibrium excistence theorem without complete or transitive preferences. *Journal of Mathematical Economics*, 1:237–46, 1974.

[136] A. Mas-Colell. *The Theory of General Economic Equilibrium: A Differentiable Approach*. Econometric Society Monographs. Cambridge University Press, Cambridge, 1985.

[137] J.M. Masque and A. Valdés. Characterizing the Blaschke connection. *Differential Geometry and its Applications*, 11(3):237–43, 1999.

[138] M.A. Maurice. *Compact Ordered Spaces*. Princeton, 1965.

[139] J. Michelsen. Preference and demand functions. Master's thesis, for math. econ. degree, University of Copenhagen, October 1997.

[140] L. Nachbin. *Topology and Order*. Princeton, 1965.

[141] J. von Neumann and O. Morgenstern. *The Theory of Games and Economic Behavior*. Princeton, first edition, 1944. second edition 1946.

[142] J. Neyman and E.S. Pearson. Contributions to the testing of statistical hypothesis. *Stat. Res. Mem.*, 1936,38.

[143] E.A. Ok. Expected utility without the completeness axiom. Department of Economics, New York University, January 2001.

[144] K.R. Parthasarathy. *Probability Measures on Metric Spaces*. New York, 1967.

[145] A.L. Peressini. *Ordered Topological Vector Spaces.* New York, 1967.

[146] J. Pfanzagl. *Die axiomatische Grundlagen einer allgemeinen Theorie des Messens.* Würzburg, 1959.

[147] J. Pfanzagl. *Theory of Measurement.* Würzburg, 1968.

[148] G. Pickert. *Projektive Ebene.* Berlin, 1968.

[149] L.S. Pontryagin, V.G. Boltyanskii, R.V. Gamkrelidze, and E.F. Mischenko. *The Mathematical Theory of Optimal Processes.* Interscience, New York, 1962.

[150] J.K.-H. Quah. Weak axiomatic demand theory. Oxford, March 13 2000.

[151] F. Radó. Equations fonctionelles caracterisant les nomogrammes avec trois echelles rectilignes. *Mathematica (Cluj)*, 1:143–66, 1959.

[152] F. Radó. Sur quelques equations fonctionelles avec plusieurs fontions a deux variables. *Mathematica (Cluj)*, 1:321–39, 1959.

[153] F. Radó. Eine Bedingung für die Regularität der Gewebe. *Mathematica (Cluj)*, 2:325–34, 1960.

[154] F.P. Ramsey. *Truth and Probability.* Paul, Trench, Trubner, and Co., 1931.

[155] K. Reidemeister. Topologische Fragen der Differentialgeometrie. 5. Gewebe und Gruppen. *Math. Zeitschrift*, 29:427–35, 1929.

[156] L. Rigotti and C. Shannon. Uncertainty and risk in financial markets. Berkeley, April 2002.

[157] A. Robertson and W. Robertson. *Topological Vector Spaces.* Cambridge, 1964.

[158] P. de Saint-Robert. De la résolution de certaines équation a trois variable par la moyen d'une règle glissante. *Mem. Accad. Sci. Torino*, 25:53–62, 1871.

[159] P.A. Samuelson. *A Long-Open Question on Utility and Conserved-Energy Functions*, chapter 15 in Essays in Honor of David Gale edited by Mukul Majumdar, pages 287–306. St. Martin's Press, 1992.

[160] L.J. Savage. *The Foundation of Statistics.* Wiley, 1954.

[161] H.H. Schaeffer. *Topological Vector Spaces.* New York, 1966.

[162] G. Shafer. *A Mathematical Theory of Evidence.* Princeton, 1976.

[163] W.J. Shafer. The non-transitive consumer. *Econometrica*, 42:913–19, 1974.

[164] W.J. Shafer and H.F. Sonnenschein. Equilibrium in abstract economies without ordered preferences. *Journal of Mathematical Economics*, 2:345–48, 1975.

[165] L. Shapley and M. Baucells. Multiperson utility. WP UCLA, July 17 1998. Part of the paper is based on an earlier unpublished paper by Shapley.

[166] W. Sierpinski. *Cardinal and Ordinal Numbers.* Warsaw, 1958.

[167] C. Skiadas. Conditioning and aggregation of preferences. *Econometrica*, 65(2):347–67, March 1997.

[168] C. Skiadas. Subjective probability under additive aggegation of conditional preferences. *Journal of Economic Theory*, 76(2):242–71, October 1997.

[169] H. Sonnenschein. *Demand Theory Without Transitive Preferences, with Applications to the Theory of Competitive Equilibrium*, chapter 10, pages 215–23. Harcourt-Brace-Jovanovich, 1971.

[170] C. Starmer. Developments in non-expected utility theory: The hunt for a descriptive theory of choice under risk. *Journal of Economic Literature*, XXXVIII(2):332–82, June 2000.

[171] V. Strassen. The existence of probability measures with given marginals. *Annals of Mathematical Statistics*, 36:423–39, 1965.

[172] W. Szmielev. *From Affine to Euclidean Geometry*. Warsaw, 1983.

[173] G. Thomsen. Un teorema topologico sulle sciere di curve e una caratterizzazione geometrica delle superficie isotermo-asintotiche. *Bollettino della Unione Matematica Italiana*, 6:80–85, 1927.

[174] G. Thomsen. Schnittpunktsätze in ebenen Geweben. *Abh. Math. Sem Univ. Hamburg*, pages 99–106, 1929.

[175] C. Villegas. On quantitative probability sigma-algebras. *Annals of Math. Stat., 35*, pages 1789–1800, 1964.

[176] C. Villegas. On qualitative probability. *American Mathematical Monthly*, 74:661–69, 1967.

[177] K. Vind. *Mean Groupoids*. Institute of Economics, University of Copenhagen, 1969.

[178] K. Vind. Additive utility functions and other special functions in economic theory. WP90-21 Institute of Economics, Copenhagen, 1990.

[179] K. Vind. Figure 4. Discussion Paper A-341 Bonn, July 1991.

[180] K. Vind. Independent preferences. *Journal of Mathematical Economics*, 20(1):119–35, 1991.

[181] K. Vind. von Neumann Morgenstern preferences. *Journal of Mathematical Economics*, 33:109–122, February 2000.

[182] P. Wakker. References compiled by Peter Wakker. http://www.fee.uva.nl/creed/wakker/refs/rfrncs.htm, March 16 2002.

[183] P. Walley. *Statistical Reasoning*. Chapman and Hall, London, 1991.

[184] P. Walley. Towards a unified theory of imprecise probability. *International Journal of Approximate Reasoning*, 24(2-3):125–48, 2000.

[185] L.A. Wasserman. Belief functions and statistical inference. *La Revue Canadienne de Statestique*, 18(3):183–96, 1990.

[186] L.A. Wasserman. Prior envelopes based on belief functions. *The Annals of Statistics*, 18(1):454–64, 1990.

[187] L.A. Wasserman and J.B. Kadane. Symmetric upper probability. *The Annals of Statistics*, 20(4):1720–36, 1992.

[188] H. Wold. A sythesis of pure demand analysis. *Skandinavisk Aktuarietidskrift*, 26-27:85–118, 220–63; 69–120, 1943-44.

[189] H. Wold. *Demand Analysis*. New York, 1952.

[190] L. Zhou. Subjective probability with continous act spaces. *Journal of Mathematical Economics*, 32(1):121–30, August 1999.

Index